Domicide
The Global Destruction of Home

T0133309

"Their eyes see rubble, former exiles see home."
Globe and Mail, 23 June 2000

Media reports describing the destruction of people's homes, for reasons ranging from ethnic persecution to the perceived need for a new airport or highway, are all too familiar. The planned destruction of homes affects millions of people globally; places destroyed range in scale from single dwellings to entire homelands. *Domicide* tells how and why the powerful destroy homes that happen to be in the way of corporate, political, bureaucratic, and strategic projects. Too frequently, this destruction is justified as being in the public interest.

Douglas Porteous and Sandra Smith begin their analysis by examining just how important home is to human life and community. Using a multitude of case studies of displacement, they derive a theoretical framework that addresses the motives for, methods, and effects of domicide. Two case studies of resettlement resulting from hydro-electric power development in British Columbia are used to test this framework. Porteous and Smith assess the implications of loss of home, evaluate current efforts at mitigation, suggest better policies to alleviate the suffering of the dispossessed, and – as a last resort – urge resistance against unacceptable projects.

J. DOUGLAS PORTEOUS is professor of geography at the University of Victoria, British Columbia. He has published eleven books on urban planning and the environment.

SANDRA E. SMITH is adjunct assistant professor of geography at the University of Victoria, British Columbia, and works as a consultant in water management and planning.

Domicide

The Global Destruction of Home

J. Douglas Porteous
and
Sandra E. Smith

McGill-Queen's University Press
Montreal & Kingston • London • Ithaca

© McGill-Queen's University Press 2001
ISBN 0–7735–2257–3 (cloth)
ISBN 0–7735–2258–1 (paper)

Legal deposit fourth quarter 2001
Bibliothèque nationale du Québec

Printed in Canada on acid-free paper

This book has been published with the help of a grant from the
Humanities and Social Sciences Federation of Canada, using funds
provided by the Social Sciences and Humanities Research Council
of Canada.

McGill-Queen's University Press acknowledges the financial
support of the Government of Canada through the Book Publishing Industry Development Program (BPIDP) for its activities. It
also acknowledges the support of the Canada Council for the
Arts for its publishing program.

We thank the following for their kind permission to reproduce
material: for lines from Robert Frost's "The Death of the Hired
Man" from THE POETRY OF ROBERT FROST edited by Edward
Connery Lathem, copyright 1969 by Henry Holt and Co., reprinted
by permission of Henry Holt and Co., LLC; James Fenton for lines
from "German Requiem," in *The Memory of War and Children in
Exile, Poems 1968–1983*, Salamander Press 1982; Excerpt from
"One Art" from *The Complete Poems: 1927–1979* by Elisabeth
Bishop. Copyright © 1979, 1983 by Alice Helen Methfessel.
Reprinted by permission of Farrar, Straus and Giroux, LLC; and Alice
Hambleton for extracts from a private letter to author Smith. Every
effort has been made to seek permission to reproduce copyright
material. If any proper acknowledgement has not been made, we
invite copyright holders to inform us of the oversight.

National Library of Canada Cataloguing in Publication Data

Porteous, J. Douglas (John Douglas), 1943–
    Domicide: the global destruction of home

Includes bibliographical references and index.
ISBN 0–7735–2257–3 (bound).—ISBN 0–7735–2258–1 (pbk.)

1. Forced migration.   2. Relocation (Housing)   3. Home.
I. Smith, Sandra Eileen, 1942–   II. Title.

HQ518.P67 2001   304.8   C2001–900718–3

Typeset in 10/12 Sabon by True to Type

For Gavin and Jeremy

Contents

Preface

Authority does not illtreat its subjects out of indifference, venality,
incompetence, or callousness, but for the common good. However arbitrary
and cruel it may seem in its actions, it is always benign at heart ...
What disabling misconceptions about human nature and society are
inspired by such lies!

                                             Vizinczey (1986, 73)

This book is about how and why powerful people destroy the homes
of the less powerful, which happen to be in the way of corporate, political, or bureaucratic projects. The places destroyed range in scale from
a single dwelling to an ethnic homeland (e.g., in Kosovo and East
Timor, 1999). The means of destruction range from warfare through
economic development and urban renewal to the creation of roads, airports, dams, and national parks. Too frequently, the elimination of
home or homeland is justified as being in the public interest or for the
common good. Indeed, many of those unwillingly displaced from their
homes for such purposes may be considered "victims of the common
good." Across the globe, at least thirty million people are such victims,
a number equivalent to the United Nations' official total for cross-border refugees.

Currently, no word exists for the action of destroying peoples'
homes and/or expelling them from their homeland. We suggest the
neologism "domicide," the deliberate destruction of home that causes
suffering to its inhabitants. A second term, "memoricide," concerns
deliberate attempts to expunge human memory, chiefly through the
destruction of memory's physical prop, the cultural landscape (e.g., in
Muslim Bosnia, 1990s). This book, then, is about domicide, and we
will demonstrate that it is not confined to military acts; indeed, it is
common and of frequent occurrence.

The difficulty of treating this neglected and complex phenomenon
requires us to explain in some detail our book's method, structure,

audience, and voice. In our attempt to create a new concept, we have chosen to take a broadly humanistic, qualitative stance, with an emphasis on the synthesis of materials from a wide range of disciplines. Our method is to gather a large number of case study examples (approximately 200, from about seventy countries), search through these data for commonalities and differences, and only then attempt to create a conceptual framework or typology that captures the essence of domicide. This, we believe, is the core of the inductive approach to science.

This method structures the book. In the first chapter, we tentatively state the concept. The second demonstrates that home is important and that its loss may result in trauma and grief. Chapters 3 and 4 are sets of case examples structured largely by motive and which suggest tentative conclusions. These early conclusions are tested by two major case studies in chapter 5. The accumulating conclusions of the latter three chapters are then used to generate a typology of domicide. Finally, we critique current modes of ameliorating the problem and suggest more innovative methods.

Because the issue is clearly a practical as well as a scholarly one, we aim for multiple audiences. We hope the book will be of value to scholars in social science, planning, and administration, as well as to graduate and senior undergraduate students – our future citizens, scholars, and policy-makers. Further, we hope the book will be read also by both policy-makers and by lay people, especially, among the latter, the victims of the common good.

The audience dictates the style used; it is multiplex. The book is co-authored by Doug Porteous – an academic with an interest in planning critique and a radical-humanist bent – and Sandra Smith – an academic and former civil servant with many years of experience in both resource and urban policy-making. Our work, however, is not dual in tone and voice. Indeed, we have adopted three stylistic voices: academic, policy-making, and one we term "informal", which is common in serious non-fiction. The chapters follow a pattern, which begins with all three voices (Chapter 1), moving to academic (Chapter 2), then informal/academic (Chapters 3 and 4), and academic/policy-making (Chapters 5 and 6), and finally returning to all three blended voices (Chapter 7). The academic style is the core style throughout.

These scholarly and stylistic approaches, we believe, are the most appropriate means of telling the story.

Acknowledgments

We'd like to acknowledge the assistance, insights, comments, and support of Judith Allen, anonymous reviewers, anonymous informants, Ian Baird, Bizet, Susan Bullock, Nandan Divakaran, Shirley Embra, Harry Foster, Katharina Ganz, Alice Hambleton, Georgina Henderson, Jill Jahansoozi (who also did much of the typing), Michael Jones and Venka Olsen, Mick Jagger and The Rolling Stones, Lisa Kadonaga, John Lennon, Maeve Lydon, Beryl-Anne Massey, Roselee Miller, Catherine Milsum, Barbara Parker, Carol and Gavin Porteous, Lloyd and Ruth Sharpe and Pamela Turyk, Peter and Jeremy Smith, Peter Smith (University of Alberta), Neil Swainson, Jim Wilson, and Janet Wood and Henry Wiseman, as well as the residents of Rapanui (Easter Island) Chile, Howdendyke, East Yorkshire, and the Columbia Valley, British Columbia. We are also grateful to the BC Government Archives (Fran Gundry, Katherine Henderson), BC Hydro (Al Geissler, Tim Newton), the Islands Trust Council and staff, the BC Ministry of Environment, Lands and Parks and its Water Management Program and the University of Victoria Archives (Chris Petter). We've also learned much from the many students with whom we have discussed our ideas. Finally, we thank the Social Sciences and Humanities Research Council of Canada for research grants, Ron and Adrianna Edwards of Focus Strategic Communications for editing, and Philip Cercone, Joan McGilvray, and staff at McGill-Queen's Press for bringing our manuscript to fruition.

DOMICIDE

CHAPTER ONE

Introducing Domicide

Someone has to lose, said the stranger. That's economics. The question is –
who loses? That's progress.

Winifred Holtby (1936, 95)

"A man's home is his castle," runs the old sexist adage. Many of us act
out this aphorism on a daily basis as we leave home's warm comfort
for the daily round. Then, at the end of the day, we gladly re-enter our
homes – places that are quiet refuges from the outside world; places in
which we can truly be ourselves and display and nurture our being;
places in which, above all, we may experience centredness, identity,
and security.

The security of our home, however, is never completely inviolable.

Moth and rust may corrupt benignly within, but when thieves break
through and steal, we often experience a feeling of violation. But
home invaders and burglars are only the most obvious threats to a
home's peace and security. As yet, few of us are subject to official surveillance or search warrants. A far more serious threat to the home is
the desire of governments or businesses to raze it and erect something
quite different on its site. Citizens may lose their dwellings through
expropriation – the power of compulsory purchase – for the common
good or in the public interest. Where the loved dwelling – or, more
likely, the cherished neighbourhood or landscape – once stood, there
is now a park, an airport, a highway, a reservoir, or perhaps a rubble-
strewn wasteland awaiting development. This deliberate destruction
of home against the will of the home dweller, we call *domicide*. Briefly
domicide is the murder of home. It is important, initially, to ascertain
whether domicide is meaningful and whether or not it is of common
occurrence.

DOMICIDE IS MEANINGFUL

People are attached to places as they are attached to families and friends.
When these loyalties come together, one then has the most tenacious cement
possible for human society.

Lewis Mumford (1961, 287)

The meaningfulness of domicide resides in the probability that home is
central to our lives, and the likelihood that the forcible destruction of
it by powerful authorities will result in suffering on the part of the
home dweller. These possibilities are approached via both anecdote
and academic literature.

In July 1993, the inhabitants of Quesnel, British Columbia,
"watched in horror as the wooden house was smashed to pieces by a
bulldozer" (*Victoria Times-Colonist*, 1 September 1993). Estranged
from his common-law wife Mildred, Leon Hetu found the best way,
short of grievous bodily harm, to do her the greatest possible economic and psychological damage. After he had demolished her house,
neighbours were able to recover only some clothes, a few splintered
chairs, and a few photographs for the dazed Mildred. Had her family
photographs been destroyed as well, Leon would have had the satisfaction of committing *memoricide* (Wilkes 1992). For his crime of
domicide, he was charged only with "mischief."

A much-better-known case of wilful domicide is Graham Greene's
short story *The Destructors*, a frightening tale of how a gang of boys
methodically tear apart a lovely period house that lies temporarily
vacant. Using hammers, chisels, and saws, the vandals rip out the fine
panelling, the parquet floors, and the handsome old staircase. When
the owner returns, the boys waylay him, lock him in an outhouse, and
continue their work until even the foundations of the house are weakened beyond repair. Tying one of the few remaining supports to the
back of a nearby truck, the boys watch as the unsuspecting driver
pulls down the house, scattering rubble everywhere. When the unfortunate owner is released and surveys the devastation, he weeps bro-
ken-heartedly as he tries to comprehend why anyone would ruin his
home so deliberately.

This story was more than just a work of fiction. The sardonic
Greene apparently wanted his readers to see a comic element in the
story. Yet on its publication in 1954, London's *Picture Post* was over-
whelmed by thousands of letters from readers offended by the graphic
cruelty of this act of wanton domicide. But Greene had a deeper
motive. Legally separated from his wife, Vivien, he knew that she had
a strong sentimental attachment to their former home. Vivien was in

no doubt about the story's meaning: "It was to me very cruel. It was as if he enjoyed the destruction. I could recognise that it was my house, our house" (Shelden 1994, 12). Ever the betrayer (Porteous 1990), Greene was striking a blow that he could be sure his wife would feel very deeply.

The wilful destruction of a loved home can thus be one of the deepest wounds to one's identity and self-esteem, for both of these props to sanity reside in part in objects and structures that we cherish. If the house has been built or restored by the dweller and the surroundings lovingly shaped, the pain will be much worse. But one's house is much more likely to fall victim to government fiat than to an angry lover. And when this occurs, mental anguish is accompanied by bewilderment, for we are invariably told that the destruction of our home is in the public interest and that our loss is a contribution to the common good.

In July 1991, such anguish took the form of a shootout, which was shown on BBC television news. Albert Dryden fired into a crowd of about twenty people who were standing outside the home he had built himself in Bustfield, northern England. Symbolically, one of his bullets killed Harry Collinson, the local director of planning. For three years, Dryden had battled Collinson and the local planning authorities who claimed that he had built his dwelling without planning permission. His last legal appeal exhausted, he was moved to the act of violence at the sight of local authority officials approaching his home with a bulldozer. His neighbours are equivocal about the now-jailed Dryden's extremism, but some regard him as a martyr regarding the rights of private property and the sanctity of the home.

An even more chilling case of resistance to home destruction at a much wider spatial level occurred a few years ago in Lesotho, a small nation that is an enclave surrounded entirely by the Republic of South Africa. Perennially short of water, South Africa looks to Lesotho's mountain valleys as an endless source of water and hydroelectric power. The World Bank agreed to fund the Highlands Water Project, which would involve the building of several dams and the flooding of some of the few valleys containing Lesotho's scarce agricultural land. Although various groups battled the vision of Lesotho as a mere resource adjunct to South Africa, one local diviner took a more traditional approach. A surveyor's helper working on the hydro scheme was found murdered in the hills; the body had been split from throat to crotch, the entrails had been arranged in neat patterns on the grass, the heart and genitalia had been removed for ritual purposes, and the eyes had been removed and reinserted backwards (De Villiers and Hirtle 1997, 114). We flinch in horror at such witchcraft, or would perhaps like to substitute a World Bank official as the sacrifice. On reflection, however, we may come to see this action as a desperate traditionalist attempt to counter the overwhelming power of detribalized, Western-educated government development "experts" who so often pursue the national interest at the expense of local needs. Moreover, while we are appalled at the barbarity of this homicide, we often fail to react to the far greater barbarity of destroying whole valleys, homelands, and cultures.

The moral of these stories is that place is meaningful to people, and that the place called home is the most meaningful of all. When it is threatened, we are roused to defend it. We also learn that home is not simply one's dwelling, but can also be one's homeland or native region. It is one of the obvious facts of life, so often overlooked, that people are not merely attached to other people but also to familiar objects, structures, and environments that nurture the self, support the continuity of life, and act as props to memory and identity. The theme of attachment is a common one in psychiatry; we have little trouble understanding the human need to be connected to others. But the theme of human attachment to place has received much less consideration. Yet geographers have investigated this concept for generations and more recently have been joined by environmental psychologists, sociologists, architects, and planners in confirming the importance of the human-environment connection (Porteous 1977). But so common is our pragmatic, instrumentalist, economist's dreary view of the human-environment relationship that "intangibles" such as attachment, aesthetics, ethics, and spirituality (Porteous 1996) are given short shrift. They cannot, for example, be honestly incorporated into the heartless cost-benefit analyses that so often determine what is to be destroyed in the pursuit of "progress."

Yet by the 1990s, the theme of human attachment to place – and especially to home – had developed a considerable academic and popular non-fictional literature. On the popular side, a British anthology of prose and poetry, *Home Is Where the Heart Is* (Anonymous 1997), is matched by Gallagher's *The Power of Place: How Our Surroundings Shape Our Thoughts, Emotions and Actions* (1993) in which many American examples pay particular attention to home. The very sensitive book by Nora Johnson (1982, 7) *You Can Go Home Again*, demonstrates that "To Americans, the very term *home* is intense and reverberating, thick with images and dreams," and that "we invest the places where we live with a lifetime of images about *home*."

In academia, geographers, anthropologists, designers, and planners frequently get together in conferences about place attachment with both theoretical and practical ends in view. I (Porteous) attended two such conferences in 1992. One in Marmaras, Greece, considered whether the "sense of place" in design had changed from a simple how-to-do-it approach to an "imperative that offers making place" as a universal nostrum for global anomie (Riley 1992). Another in Trondheim, Norway, discussed the commonalities between ancient homes and modern internationalized homes and sought lessons from the past for use in the present (Benjamin 1995).

At the theoretical level, geographers have suggested that people acquire information about locations and store it efficiently in their memories in a hierarchical structure. Central to this form of organizing space is the personal hierarchy: my home; home neighbourhood; home city, home region, and home country (Lloyd et al. 1996). This centredness of home is felt most keenly when home is lost: "To be forcibly evicted from one's home and neighbourhood is to be stripped of a sheathing which in its familiarity protects the human being from the outside world" (Tuan 1974, 99).

At the practical level, we might ponder the amount of local effort taken to ensure the preservation of the tiny Norwegian neighbourhood of Ilsvikøra close to the centre of Trondheim. Surrounded by factories and docks, the community of twenty-seven wooden houses built around 1860 is one of the few intact nineteenth-century working-class environments left in Trondheim. Threatened with domicide by industrial expansion, residents combined with local academics to mount an exhibition of photographs and oral-history tape recordings and drew up a rehabilitation plan in 1972. By 1980, the district had been saved, and its success encouraged other projects such as the one at Footdee in Aberdeen, Scotland (Jones and Olsen 1977). Ilsvikøra is now a pleasant home neighbourhood, an unexpected joy for the observant visitor to Trondheim's dockland.

DOMICIDE IS COMMON

It has been said that there is not one corner of the planet ... that has not been considered by someone to be the most beautiful place on earth. That place is their place, a place they call home. But for millions of people across the globe, home is now a lost memory, or a dream. These are people who have lost their place ... The dislocated, the displaced, the homeless.

Inter Pares, *Bulletin* (1994, 1)

This book is about the frequency of domicide. The following brief account draws from literature, history, and international statistics. Historical accounts often pass over domicide, with all its attendant misery, in a single phrase: the town was burned; three hundred villages were destroyed; the whole country was laid waste. Certain countries such as Ireland and Scotland have a considerable literature on the "clearances" by the English which had native Celt tenant farmers uprooted and replaced by sheep. To take only one short period, the England of Henry VIII has been called the "Age of Plunder" (Hoskins 1976). Accurately depicted in Holbein's famous portrait as a voracious, merciless, porcine predator, Henry was a leader in the conspiracy of the rich to become yet richer by dispossessing the poor of their homes and lands for lucrative sheep farming. Like some "leaders" today, Henry was an efficient kleptocrat, who put to death those who, like Sir Thomas More, objected to this and other injustices.

Such landscapes of cruelty are to be found throughout history, and in every part of the world today. In fact, the twentieth century has been called the century of the "displaced person." Eleven millions of these, who the vicious victor General Patton dismissed as animals, wandered through central Europe in 1945; this number was surpassed two years later when fifteen million people were uprooted at the partition of India. Like British eighteenth-century poetry at the time of the "enclosures," Third World novels are replete with images of domicide. Even modern British comic novels, soap operas, and detective stories use domicide as a theme. To take only a single example, in P.D. James's *Innocent Blood* (1989, 181), we find members of a London neighbourhood whose "chief preoccupation was the rumour and threat of a local authority development which would sweep their world away."

Nor can the world traveller escape encounters with cases of domicide. Three recent guide books to Southeast Asia casually note a variety of incidents. In *The Vietnam Guidebook* (Cohen 1990, 352), the section on Ho Chi Minh City (Saigon) describes how the citizens of the exurban area were removed via the Strategic Hamlet Program so that American troops could bomb and strafe the Viet Cong's Cu Chi tunnel complex. Angry residents found themselves relocated to fortified villages, while their former farmland was defoliated and resown with "American grasses," which could rapidly be burned off to expose enemy positions. Some citizens were moved by this outrage to join the guerillas, extend the tunnel system, and further prosecute the war. The Cu Chi tunnels are now tourist attractions; here, former American shell cases and bullets are fashioned into souvenirs and sold to visitors.

The *Philippines Handbook* (Harper and Fullerton 1994, 181) notes that the Cordillera Central of Northern Luzon is internationally known for the "native opposition" to the proposed Chico River Dam project, which was later suspended. And a regional handbook (Eliot et al. 1993, 476) explains to visitors to the temple city of Pagan, northern Burma, that "to create a pristine, peasant-free, historical environment for tourists, and for 'archaeological reasons,' Pagan's 6,000 residents were given one week's notice and then forcibly relocated to New Pagan, a soulless and treeless wilderness 5 km south of the old city." The residents were lent one government truck for one week, had all utilities cut off, and were moved at gunpoint. Those who objected were jailed. The new site is characterized by poor land, erratic utilities, lack of community feeling, and no jobs. (One wonders if the militarists who currently run Burma impound this tourist guidebook at the airport.)

To count the numbers displaced across the world by similar projects in peace and war would be a difficult task especially because, as yet, the concept of domicide is not widely accepted. The international statistics that most closely relate to domicide have been produced by the United States Committee for Refugees, the UN Secretariat on Internally Displaced Persons, and the Worldwatch Institute. The former, which normally takes into consideration only international cross-border cases, has now created a category of "invisible refugees," those made homeless within their own countries and thus not included in official refugee figures (*New Internationalist* September 1991, 18). Their estimate for 1990 was more than thirty million people, including over four million in each of the Sudan and South Africa. No indication is given of the causes of these internal displacements, but much will be due to war and famine. Similarly, UN representatives, writing in the journal *Foreign Affairs* (Cohen and Deng 1998, 12), contended that "the newest global crisis" is "internal displacement" in which "tens of millions of people have been forced from their homes during the past decade by armed conflict, internal strife, and systematic violations of human rights, all the while remaining within the borders of their countries." However, in its *State of the World* report for 1997 (Brown 1997, 125), the Worldwatch Institute states that "over the past decade, as many as 90 million people may have lost their homes to dams, roads, and other development projects." Most of these will have experienced domicide.

The numbers, then, are indeed large, but difficult to confirm and likely to be inaccurate. Nevertheless, we expect that the amount of people currently displaced from their homes by domicide will easily exceed the twenty-five to thirty million "official" cross-border refugees recognized by the United Nations in the 1990s.

WHAT IS DOMICIDE?

It's fascinating how often the simplest observations are not made until
someone provides a conceptual framework.

R.M. Restak (1979, 205)

Domicide is very clearly a major global problem. It is often a devastating experience to its victims, although it is apparently of little concern
to its perpetrators. However, "domicide" as both a word and a concept
hardly yet exists. There are hundreds of anecdotes and academic studies of individual cases of domicide, but there seems to be a marked
reluctance to investigate similarities across cases and to construct generalizations. "Only connect," cried E.M. Forster, a radical suggestion
currently ignored by specialists who devote their attentions to dam
building or slum clearance alone.

Very few attempts have been made to generalize about the deliberate destruction of home. In his pioneering *The Language of Cities*,
Abrams (1971) comments on the negative effects upon poor families of
relocation and urban renewal. The book also has useful discussions on
the topics of condemnation, compulsory purchase, eminent domain,
eviction, and expropriation. The earliest coherent attempts were made
by Fried (1966) and Marris (1974), who placed the problems caused
by forced removal within the general category of "loss and change"
and likened the mourning behaviour of some relocatees to the grief
that follows bereavement. In concentrating on the effects of forced loss
of home, however, Marris explicitly chose to ignore the political and
economic causes of loss, an issue examined in this book.

A few years later, Gallaher and Padfield (1980) edited a collection of
essays entitled *The Dying Community* in which a number of causes
of community decline were discussed: the violent actions of nature or
of humans; the abandonment of a natural region as when a mine is
worked out; the decay of a socio-cultural system; and the global extinction of some form of association such as a trading system. After establishing the by-no-means accepted fact that settlements do die, the
authors demonstrate that small, isolated communities are especially
vulnerable to extinction and that the decision to kill such a place is
usually made by some distant authority in a corporate boardroom
or government office. The book also stresses the incompleteness of
decline; many small places experience "negative growth," but few disappear completely. There is persistence in the midst of decay. It is
because the authors concentrate on small, isolated settlements and fail
to adequately consider urban slum clearance or wilderness dambuild-
ing that they provide an inadequate account of the process and effects

of domicide. This book will hopefully demonstrate that many places do not merely decline but are, in fact, killed and often disappear entirely. Nevertheless, the Gallaher and Padfield book points the way to domicide and accepts the notion that community decline may be a source of personal grief.

Simultaneously, Coelho and Ahmed (1980) approached the problem, in *Uprooting and Development*, from the viewpoint of mental health. In this large book that covers the problems of foreign students, new settlers, immigrants, children, and adolescents, only Trimble's chapter on forced migration has direct relevance to the concept of domicide. In a study of the coerced relocation of four indigenous groups, Trimble shows how these groups were denied decision-making power and how the goals of the agencies that perpetrated the removals were incompatible with the goals of those forced from their homes. Trimble (1980, 475) concludes that: "Planners should be more concerned about the effects of abrupt social change on a community," and "forced relocation of culturally diverse groups is a topic that has received far too little interest." In the same volume, Marris (1980, 114) notes that: "Again and again, actions, seemingly designed to help people, frustrate and bewilder them by alienating them from the context of their lives as they perceive it." This is the crux of the domicide issue; what is uprooted may be the very meaning of people's lives.

This was certainly the case during the massive aerial bombardments of World War II, researched in some detail by Hewitt (1983b). In a very sympathetic consideration of these horrors, Hewitt suggests that just as biologists have prepared "red books" of endangered species and ecologists have produced "green books" of threatened habitats, so we also need "black books" of places destroyed or almost destroyed by human violence. In the first attempt to create a neologism for this process, Hewitt suggests "place annihilation."

All of these views are partial ones. First, some concentrate on the dying of places, whereas others insist that places can be killed. Second, they each tend to consider only one or two terrains of destruction such as inner cities, small rural settlements, or the island and forest habitats of indigenous peoples. Third, each also concentrates on only a few processes of destruction: corporate economic change, government urban policies, and wartime bombing. What is needed is a comprehensive, holistic framework that covers all processes of deliberate home destruction in all types of landscape. That concept is domicide.

Domicide is a neologism that first appeared in a series of books, articles, and speeches by Porteous (1988, 1989). It is not yet found in dictionaries, in which it may appear one day, very appropriately between "domesticity" and "domicile." The central concept is home (Latin:

domus), and the issue is deliberate home destruction. The concept of home is important and amazingly complex, taking up many pages in the larger dictionaries. The suffix "-cide" indicates not merely death, but deliberate killing, as in homicide and suicide. Briefly, then, domicide is the deliberate killing of home.

Concepts relating to domicide include eviction, exile, expropriation, displacement, dislocation, and relocation. In French, we would consider *déracinement* (uprooting), which for the victims may lead to *dépaysement* (a feeling of strangeness and disorientation) or even *déclassement* (a relegation to lower status). All of these words are problematic since most of them are partial and exclusive. It is significant that in Stoett's (1999) exploration of terminology related to human and global security, he is able to conceptualize genocide, ecocide, and globalization in single nouns that denote meaningful process. However, a chapter that might well be labelled "domicide," is entitled "Population Displacement: Refugees," is a neologism seems necessary.

The first attempt at creating a suitable neologism was "topocide," the killing of place. But it was rejected because of its ungainly mixture of Greek and Latin and because of the vagueness of meaning of the word "place." A second attempt generated both "topocide" and "domicide" as a paired antinomy. The first was to denote the destruction of home from the point of view of the perpetrators – normally outsiders – and would emphasize motive and process. The second term would involve the reactions and responses of the victims – always insiders – and thus would emphasize effects. Despite the neatness of this system of juxtaposing the dichotomies of process-effect, outsider-insider, and perpetrator-victim, the formulation was judged to be too complex. Far better, then, to include all these components in the single new term: "domicide," the killing of home. More formally, domicide is defined as *the deliberate destruction of home by human agency in pursuit of specified goals, which causes suffering to the victims*. In addition, we specify that the human agency is usually external to the home area, that some form of planning is often involved, and that the rhetoric of public interest or common good is frequently used by the perpetrators. It follows that home destruction perpetrated by or welcomed by the home dweller cannot be domicide; the notion of suffering is crucial.

As these are controversial terms, it is worthwhile to pursue the concepts of "victim" and "common good" in some detail if only because victimhood may be recognized but dismissed by project proponents because of the immeasurably greater benefits they believe they are bringing to the public at large. In seeking a modern definition of victimhood, Weisstub (1986, 317) correctly points out the long history of victimization beginning with the Garden of Eden and, we hope, ending with the twentieth century's victims of holocaust and genocide. He also recognizes that scientific detachment was possible only with the historical perspective and a focus on groups rather than individual victims. Such freedom from bias becomes less easy and the definition of victim correspondingly more difficult in contemporary times given the realization that there are victims and aggressors in political, economic, familial, and emotional life and given the modern media's role in exposing the "plight" of the victim and the "evil" of the aggressor. Nevertheless, the definition of persons who lose their homes through domicide as victims seems justifiable given Weisstub's description of a victim as "a person who has been unjustly treated" and whose "human or economic power has been weakened."

There is a connection between the notion of the domicide victim and what may be the real aggressor in domicide, the oft-invoked concept of the common good. The term common good is almost always defined by the elite or the majority and is used interchangeably with the term "public interest." Raskin (1986, 38) suggests a long history for the notion of common good – from Thomas Aquinas who believed that profit making would benefit chiefly the powerful individual or corporate organization, which would then act against society itself, to well-founded beliefs that point out the contradictions between capitalism and the pursuit of the common good. More recently, Massam (1999, 347) has provided a good short introduction to the similar notion of "the public good" and believes that "the topic of the public good provides a challenge to the next generation." Yet while acknowledging that the public good is being undermined by commodification, corporatism, and individualism, he does not significantly question the concept itself. Fortunately, Fagence (1977, 83) provides an omnibus definition of public interest in relation to planning: "The public interest is promoted or protected if the community is able to enjoy increased or improved facilities, amenities and services; if the provisions are sufficient (quantity) and adequate (quality); if they are convenient, efficient, compatible, not exclusive, free of onerous restrictions; if minority interests are wholly recognized and accommodated; if external (geo-political) relationships are not prejudiced; and if most other individual rights and privileges are not unduly constrained or denied." In this book, in which loss of home is a central theme, individual rights are of significant interest. These individual rights include both the rights claimed by owners of private property and the much-less-tangible rights associated with the creation and enjoyment of a home. Thus, from the point of view of domicide, the common good concept is flawed, for it excludes the victims it creates.

WHAT DOMICIDE IS NOT

It is also useful to distinguish domicide from a number of related issues that may shed light on some of the concepts inherent in domicide. Domicide is a lesser horror than *genocide*. While domicide requires that victims remain alive to suffer the loss of home and perhaps rebuild their lives, genocide requires the deaths of the victims. Some of the earliest accounts of genocide are to be found in the earlier sacred books of Mesopotamians, Jews, and Christians. The commandment "Thou shalt not kill" clearly did not apply to those already inhabiting the land flowing with milk and honey; they were not only to be dispossessed but also slaughtered. The Hebrews began by killing the men, women, and children of the Midianites and the Amorites as well as the subjects of Og, the king of Bashan (Numbers 31–2, Deuteronomy 1–3). Worse was to befall the Canaanites; the book of Joshua is a record of genocide committed upon over thirty small kingdoms. Not only were all the people killed, but their animals were slaughtered and their towns burned until only rubble remained. The feeble excuse for such genocide, as given by biblical commentators, is that the issue is not edification but obedience, and that good can come out of evil.

This brings us to the issue of who benefits from genocide and domicide, and who loses. The Canaanites clearly suffered a great evil at the hands of the Jews, who also tried to commit memoricide by casting down the religious structures of Canaan. The Romans behaved slightly better in AD 70 and 136, destroying Jerusalem but permitting most of the Jews to seek exile – domicide rather than genocide. Hitler's aim was clearly genocide, and the surviving Jewish remnant that returned to Poland in 1945–46 only to see their homes and city quarters razed to the ground, suffered further deaths at the hands of Poles and Lithuanians. Little wonder that, memory intact after two millennia, Jews flooded into Palestine after 1945 and committed domicide against about one million Palestinians in 1948 and 1967.

Since the 1960s, we have been afflicted with a wave of neologisms involving the suffix "-cide." Some of these are exotic ecospeak including the truly awful puns "countrycide," "rivercide," and "seacide." Nevertheless, these dreadful locutions do point to the fact that global environmental ignorance and greed are beginning to destroy the planet. We therefore accept the neologisms of "ecocide," the killing of an ecosystem (now routine in Canada, Brazil, and Indonesia) and "terracide," the killing of the planet. On a scale of horror, domicide pales beside terracide and is often included in ecocide and genocide. But domicide has a special trauma, because the victims are not killed but must watch their homes being destroyed as they are wrenched forcibly from them followed by an attempt to overcome relocation trauma and to build a new life.

Those who escape such devastations are called homeless exiles, or refugees. Domicide differs from all three of these vague concepts. Homelessness, first, is generally conceived of as roofless people sleeping on the streets (Kearns and Smith 1994). The victims of domicide usually obtain a new roof; their issue is that they preferred the old one. While homelessness in the Third World is rarely studied, a large literature is building up on it in the Western world. One reason for this is that the West's homeless are small in numbers but obvious in location. Their cardboard shelters occupy the business and shopping districts of city centres. Their bodies are living witnesses to the inadequacies of the capitalists who make deals in the glittering office towers that rise above them.

Nor are domicide victims to be confused with those suffering exile. Since Ovid, exiles have been mostly political, often literary, intellectuals (Simpson 1995). They may choose their way of life, and, for many, theirs is "a dream of glorious return" (Rushdie 1988, 505). If home is destroyed, as in domicide, returning becomes problematic. And in general, we may conclude that the exile is a rather privileged person; the underprivileged are called refugees.

International refugees have much in common with the victims of domicide particularly in terms of forced movement and relocation (Rogge 1987, Black and Robinson 1993). Numbers are difficult to assess, but sources such as the United States Committee on Refugees and the United Nations High Commission on Refugees (UNHCR) suggest seventeen million in 1989, rising to twenty-seven million by 1995 (Van der Gaag 1996). Much of this rise has been due to political changes in the former Soviet Union. According to the UNHCR, more than nine million people have abandoned their homes in the former USSR since 1989 because of ethnic tensions or ecological disaster. Almost all were forced migrants; 3.6 million fled from ethnic war zones such as Chechnya and Azerbaijan; 1.2 million returned from Stalin's deportations of 1944; and 700,000 escaped from ecological disasters such as Chernobyl, the nuclear testing zone of Semipalatinsk, and the drying up of the Aral Sea. Such movements have recently been termed "ecopolitical displacement" (Stoett 1999, 94).

Using data from the United States Department of State, Wood (1994) considers "forced migration" only in terms of countries in which over 100,000 people have been forcibly displaced and in which these numbers exceed 1 per cent of the country's total population. Nevertheless, the study estimates that thirty-three nations met these criteria in the mid-1990s, accounting for a total of 17.6 million cross-border refugees. Most of the people involved were very poor, emanated from countries in Africa and Asia, and had fled scenes of ethnic, religious, or tribal conflict.

There are two problems, at least, in considering domicide victims as refugees. First, most refugees are fleeing warfare or environmental degradation; only the former of these causes may be at all likely to involve domicide. Second, only those displaced persons who actually cross an international boundary can be registered as refugees by the UNHCR and thus receive UN benefits. Most domicide victims, in contrast, will stay within their own country. Derived from the same thirty-three countries as the refugee study, Wood's figure for such "internally displaced persons," albeit conjectural, is 20.9 million persons. Only some of these will be victims of domicide.

Wood notes that among the myriad reasons for forced migration are government-sponsored development schemes that relocate indigenous groups, including 3.6 million blacks in apartheid South Africa. Such "development" refugees are generally regarded as little more than a nuisance by governments, which ignore the profoundly disruptive changes that forced relocations cause (Oliver-Smith 1991) and usually provide inadequate compensation or other remediation for loss (Partridge 1989). Although the UNHCR has become marginally involved in the politically charged issue of protecting "internal refugees" in their own countries (Goodwin-Gill 1993), efforts are seriously hampered by sovereignty issues and funding problems (Cohen 1994). And, finally, Wood (1994, 618) laments that "Apart from a few research projects carried on by a few anthropologists, scholars have ignored people who have been forcibly displaced by government projects." It is the task of this book to remedy this myopia by focusing upon people that are internally as well as externally displaced by domicide and to look directly at the roles of the government agencies and business corporations that perpetrate such injustices.

However, injustice is not a causal issue in other types of home loss. Natural disasters such as severe weather, earthquakes, volcanic eruptions, landslides, floods, and droughts annually destroy significant numbers of dwellings and even whole settlements. Planning for the mitigation and relief of such natural disasters has received considerable attention in geography, psychology and disaster planning since the 1960s, and an enormous literature exists (White 1961, Burton et al. 1978, Wright et al. 1979, Foster 1980, Rossi et al. 1982, Hewitt 1983a, Burton 1994, Hewitt 1995). But no such literature exists for domicide. Yet according to Foster (1976), only at level XII, a catastrophic disaster, does abandonment of the site occur. Site destruction and replacement are normal in domicide.

There are significant similarities and differences between domicide and natural disasters and even perhaps some overlap since many so-called natural disasters are in fact caused by humans burning, terracing, or deforesting, as in the arson of Borneo 1997–98. In particular, severe emotional reactions follow the loss of home in both cases. After a natural disaster, however, the victims can normally blame only nature or God. Secondly, the situations are also different in that the reaction to loss from natural disasters is immediate, and, therefore, commentators have had to rely mainly on observations made soon after the event occurred. Thirdly, it appears to be endemic to our political system that such losses are mitigated as rapidly as possible, unlike planned domicide where compensation may not be given for years. Thus, recovery from natural disasters is often rapid as persons rebuild their homes in a safer way or in a more protected place.

Baum (1987) has undertaken an extensive review of the literature on the effect of natural disasters, particularly the psychological effects, which are similar to but often more severe than the results of planned domicide. He indicates that such events are the cause of social disruption, disorganization, and massive migration as well as individual reactions of trauma, fear, stress, and shock. Some studies suggest that chronic stress occurs, while others state that although the immediate psychological effect may be acute, it will also disappear rapidly. The latter circumstances may result from the immediate desire on the part of disaster victims to rebuild their homes, often supported by communities that respond with greater social cohesiveness after disasters.

Simpson-Housely and de Man (1987, 3) have also studied psychological reactions to disasters but from the perspective of how personality traits such as sense of control and anxiety affect these reactions. They confirm that knowledge of personality traits and response to natural hazards enhances understanding of the human appraisal of the hazards concerned. Such findings could also contribute to an understanding of how to respond to those who lose their homes in other circumstances. For example, certain people might be expected to act as leaders in planning a new future in which domicide must occur.

Accidental disasters, by definition, are not deliberate, but they are similar in these and other respects to natural disasters. Generally known as technological hazards, such disasters have been explored in some detail by academics (Cutter 1993), and examples such as Niagara's Love Canal are well known (Gibbs 1998). Another example of a disaster caused by technological failure is described in studies of the residents of Saunders, West Virginia, a few years after a dam failure on Buffalo Creek wiped out all traces of their town. The residents exhibited chronic symptoms of psychopathology comparable to "highly distressed" psychiatric patients (Gleser et al. 1981, 149). Their valley had been completely changed, and there was no home to which they could return. The dam failure was a human-induced error, and for those displaced, "it is ... a form of shock – a gradual realization that the community no longer exists as a source of nurturance and a part of the self has disappeared" (Erikson 1976, 302). The residents of Buffalo Creek felt that they had been betrayed by those who they normally trusted (Gleser et al.1981, 149). Only those who were able to rehabilitate still-standing homes showed less anxiety.

Perhaps the loss of home caused by the Chernobyl disaster after 26 April 1986 best exemplifies home destruction that is unplanned but also unnatural. Scherbak (1989, 4, 9) tells the story of how this short moment in time changed Chernobyl from "a pleasant little provincial Ukrainian town, swathed in green, full of cherry and apple trees" where "the beauty of Polissia nature had blended astonishingly harmoniously and inseparably with the four blocks of the power station," to an area in which devastation was totally hidden but in which the "increased radiation level would show itself in mushrooms, peat bogs, black currant bushes, and in villages at the corners of buildings where the rainwater ran from the roofs." But at first, this hidden danger was not known, and people were simply evacuated. Not knowing they were leaving their homes forever, they took with them only their summer clothes and their most important possessions. The majority of the fall-out occurred in Belarus, destroying 485 villages. One in five Belorussians, amounting to over two million people, still live in contaminated areas in which there are yearly increases in individuals suffering from cancers, genetic mutations, and neuropsychological disorders (Ignatenko 1998). Thus, lives were changed for many locally, while elsewhere, an increased distrust of nuclear power plants emerged (*Journal of Environmental Psychology*, Special Issue 1990).

Much that has been written about this aftermath has focused on finding fault and the effect on human health and ecology. However, Marples (1988, 146–7) focuses on the social impact – on people who returned to their homes and on those who did not. The first described themselves as "the happiest people in the world" because they had been allowed to regain their homes. Others, who were accommodated in apparently pleasant, permanent housing elsewhere, continued to long for their own homes. About 1,200 people, mostly retirees, eventually returned to their old homes despite the certain risk from radiation: "these people are coming back of their own volition. It's their home"; "They survived the Nazis and fear nothing" (Dodds 1989, 6). But for others, farmers and their families who had lived within 10 to 15 km of the exploded reactor, return was impossible, and their lives were changed forever. As a mark of this change, Scherbak (1989, 167) carries with him the memory of abandoned villages and particularly the village cemeteries, "'shadows of forgotten ancestors' where the living will no longer ever return."

Unplanned disasters have some similarities to domicide. Significant psychological effects are caused by the disaster event and the loss of home and community. For some, such as the elderly residents of Chernobyl and those who rebuild immediately after natural disasters, there is only one home, and that is the place they must be. Recovery is also quite rapid following a disaster event, which is not always the case for domicide. And, most fundamentally, domicide differs from unplanned disasters because the loss of home is deliberately engineered: somewhere, someone is to blame.

Great Planning Disasters (Hall 1980) is the last of the event categories which might be confused with domicide. Someone is clearly responsible in these cases, but Hall is chiefly concerned with the issue of uncertainty in planning and the economic costs entailed in poor advance estimating, as with the 1,000 per cent budget overruns for the Sydney Opera House and the Concorde. While uncertainty is a major issue for the victims of domicide, their meagre economics are generally lost within the costs of the development project as a whole. But Hall's interest in public participation in planning is directly relevant to domicide as is his conception that the major actors involved in planning disasters are the bureaucracy, politicians, private enterprise, and the local community. Bureaucrats seek aggrandizement and policy maintenance, politicians strive to retain power and enhance status, producers attempt to increase profits, and consumers try to "maximize utility." It is this unfortunate economistic phraseology that is so foreign to our studies of domicide. People are not necessarily to be considered as consumers, nor does the maximization of utility adequately capture the pleasures gained from being able to stay unmolested in one's chosen home.

In sum, domicide is the planned, deliberate destruction of home causing suffering to the dweller. Domicide victims are not normally considered exiles or refugees; they rarely cross borders and thus remain in the obscure, generally unrecognized category of the internally displaced. Nor are they the homeless, for they generally find a roof. What they have lost is their own loved home; what they have suffered is forcible removal from it for the common good. Nor are domicide victims found among those suffering from unplanned disasters. In domicide someone or some group is responsible for the suffering; we cannot blame nature, God, or even "the system." However, given the hierarchical and self-protective nature of corporate and government organizations, it may be difficult to find out who these people are and more difficult still to confront them.

Domicide is not as important an issue as terracide, ecocide, and genocide. Nevertheless, it is going on worldwide, every day, to people such as ourselves. It is immediate and it is reported frequently in the newspapers. And above all, domicide prevails where the heart is – domicide begins at home.

The remainder of this book will explain how we look at domicide.

THE ARGUMENT

No other place You could find? Here only all the trouble, always? The darkness, the flood, the fire, the fight? Why not Tata Palace? Why not Governor's mansion?

Rohinton Mistry (1993, 403)

Old Cavasji, in Mistry's *Such a Long Journey*, is as usual berating God, this time for the impending destruction of the wall between his apartment block and the street. He is berating the wrong person, for it is just those who do not suffer domicide – namely business and government people – who perpetrate it. Tata Palace and the Governor's mansion are immune. This fact, we hope, has already emerged in our discussion.

To confirm that home is extremely meaningful to its inhabitants and to stress just what is lost when domicide occurs, the concept of home at several scales is considered in detail in chapter 2. Using a wide range of social science and other literature, we conclude that home is a positive factor in people's lives since it helps to confer both centredness and identity upon both individual and group.

The following two chapters explore a multitude of cases of domicide at scales ranging from a single house to entire homelands. The information is presented in the form of many short, illustrative descriptions intended to familiarize the reader with the means of, motives for, processes, and effects of domicide. Wherever possible, the questions where, when, at what scale, why, by whom, and to whom are asked. In certain cases, which are particularly well documented such as London's Third Airport and apartheid removals, longer accounts are provided. Over 200 examples, drawn from more than seventy countries in five continents are given. They were chosen to ensure the emergence of an adequate global picture of the motives for, processes of, and results of domicide. Many of the stories remain incomplete, for to bring each one up to date would be a very time-consuming task. The information used is derived from a variety of sources, chiefly research literature, but we have also relied on the following: government reports; non-academic non-fiction; novels, poems, and plays; travellers' accounts and travel guidebooks; works of investigative journalism; and personal research and observation.

No hard and fast categorization of domicide is initially attempted. Indeed, many efforts to schematize domicide in diagrammatic form came to grief before we decided on the simple dichotomy of "extreme domicide" (generally infrequent, massive, abnormal) and "everyday domicide" (frequent, smaller-scale, "normal"). These, again, are not hermetic compartments; like acute and chronic pain, they interpenetrate considerably. Domicide is a very nuanced concept.

A number of generalizations emerge from the myriad case studies of chapters 3 and 4. To test the validity of these and to relate domicide to the meanings of home elucidated in chapter 2, our fifth chapter presents two of our own lengthy case studies of the domicide involved in hydro electric power development in British Columbia. These cases were chosen because they were accessible to us; they could be brought up to date with little difficulty; and because one author (Smith) was involved in the planning process during the 1990s. Further, they are examples of the drowning of home by reservoirs – that is, of domicide that, although of the everyday variety, is irreversible. The reader seeking a further in-depth study based on extensive fieldwork is referred to Porteous's *Planned to Death* (1989).

As we follow the route of inductive science, exploring a large number of examples of a putative phenomenon before attempting to generate a framework, typology, or theory, our generalizations on the nature of domicide are to be found in chapter 6. Besides outlining domicide's essential characteristics, we pay particular attention to the mental health outcomes of the victims as well as to the generally feeble efforts of perpetrators to provide remediation and recompense.

Finally, having confirmed that domicide is a genuine phenomenon and a major global issue in terms of both population numbers affected and the degree of injustice, we provide in chapter 7 an innovative set of approaches to its remediation, mitigation, or prevention. Those who are already convinced that home is important to human lives or who do not wish to pursue a detailed case study will find the essence of our argument in chapters 6 and 7, with supporting detail in chapters 3 and 4.

We cannot close this introductory chapter without revealing our aims and our biases. As geographers, we are aware that "geography does not amount to a mere tool for knowing about the world; it is an instrument for action" (Gilbert 1989, 222). Further, geographers have been urged to produce effective social knowledge; place an emphasis on interdisciplinary co-operation; rethink research goals as contributory to the policy-making process; and adopt a reflective spirit open to moral and human consequence (Steed 1988, 10–11). As academics, we accept Wilson's (1991, 25) urging that universities "rethink our mission [and] organize the 'scholarship of integration' which involves synthesizing results already obtained and making connections across disciplines." We also wish to counter Bok's accusation that: "Armed with the security of tenure and the time to study the world with care, professors would appear to have a unique opportunity to act as society's scouts to signal impending problems ... Yet rarely have members of the academy succeeded in discovering emerging issues and bringing them vividly to the attention of the public."

We wish, in the words of Orr (1994, 9), not to be professionals with something to sell, as academics are increasingly becoming, but persons with something to profess. In its synthesis of material from many disciplines and sources, in its moral dimension, and in its practical conclusions, this study strives to meet some of the goals set for both geographers in particular and universities in general.

We do not believe, as Orwell apparently did, that the landscapes we love are being crushed by "the relentless movement of vast, unseen historical forces" (Shelden 1991, 310). We prefer to believe with the novelist Winifred Holtby (1936, v) that the forces of change are located in human agency and are the work of powerful individuals or small groups: "The complex tangle of motives prompting public decisions, the unforseen consequences of their enactment on private lives appeared to me as part of the *unseen* pattern of the English landscape ... What fascinated me was the discovery that apparently academic and impersonal resolutions passed in a county council were daily revolutionising the lives of those men and women they affected." And our reply to Holtby's stranger (see the epigraph at the beginning of this chapter) is that his cruel economics is an amoral, competitive zero-sum game, with no possibility of a co-operative win-win situation, and further, that his conception of progress presupposes a common good rhetoric to which even the losers adhere. We do not believe this is so.

Our bias is for the victims of domicide. We believe that domicide is a moral evil. We do not agree that its victims should have their life choices so constrained and their lives so disrupted without adequate consultation or recompense as so often happens today. Many domicide events are of their nature evil, casting aside the "people in the way" in the pursuit of profit, progress, or plan. Almost all of them are fundamentally undemocratic, and their perpetrators often contemptuous of those they disempower in the public interest. As Castells (1989) explains: "People live in places, power rules through flows."

Political, bureaucratic, and business power continually flows across landscapes and overwhelms the place called home. Throughout this study, we have tried to provide the victims of domicide with a heroic stature in order that they may also be recognized. In essence, we seek to reveal the shadows that fall on the landscape when common humanity is lost in ignoring the rights and needs of others (Tuan 1993, 239) and the light that prevails in enhancing the dignity of the victims of such shadows.

CHAPTER TWO

Home:
A Landscape of the Heart

It's my favourite place, here – down the new road through the iron gate. I
stand here and watch the seasons come and go. At night, the moonlight plays
on Hunder Beck ... and the waters sing a song to me ... I know this place
will always be loyal to me. If I have nothing in my pocket, I will always have
this. They cannot take it away from me, it's mine, mine for the taking and
always will be ... even when I'm no longer here. Wherever I go ... and whatever I am ... this is me.

Hannah Hauxwell and B. Cockcroft,
Seasons of My Life, 1989

Everyone knows what "home" means. Yet, this apparently simple
concept has been the subject of countless studies, many stories, and
much art and poetry. Home has been a theme of research in disciplines
as varied as anthropology, environmental psychology, sociology,
gerontology, women's studies, history, ethnoarchaeology, architecture,
education, planning, and geography. Indeed, home is one of the central concepts of human geography. At the global scale, Carl Ritter's
geography is "the study of the earth as the home of man." At the meso
scale, Kniffen believed that mapping the types of houses in Louisiana
was an "attempt to get an areal expression of *ideas* regarding houses
– a groping toward a tangible hold on the geographic expression of
culture" (Hartshorne 1949, 230). At the micro scale, J.B. Jackson
urged that "the primary study of the human geographer must be the
dwelling ... as the microcosm, as the prime example of Man the Inhabitant's effort to re-create Heaven on Earth" (Jackson 1952, 6). Each of
these examples, together spanning the century before 1960, focuses on
physical manifestations of home yet recognizes the greater depth of
meaning.

Since the 1960s, the geographic literature on home has flourished.
For example, Mackie (1981, 7), in reviewing the roots of the study of
home, lists the following concepts: lifeworld (Buttimer 1976, Ley 1977,

Seamon 1979); attachment (Tuan 1974, 1975a, b, 1977); dwelling (Buttimer 1976, Relph 1970, Seamon 1979); rootedness (Godkin 1980, Tuan 1980); existential insideness (Relph 1976); homeland (Tuan 1974, 1977); territoriality (Porteous 1976); and home in relation to journey (Tuan 1971). To this array must be added the work of Hayward (1975, 1976) on home as an environmental and social concept and Gregson and Lowe (1995) on home and ideology. Beyond geography, many valuable studies, both humanistic and empirical-behavioural, have appeared in architectural psychology and environment and behaviour research especially those that have sought to clarify the concept of home in order to relate this concept to other variables (Rapoport 1992, 1).

The following sections contain material found primarily in the academic literature about home but augmented, on occasion, by references to fiction or poetry. This material is presented in the form of general explanations and in collections of quotations. "Concept is there ... but beyond concept is the 'concept brought into life by image'" (Brook in Bradshaw 2001, R1). The reader is invited to refer to the boxes throughout this chapter to form images of home and its meanings.

DEFINING HOME

My home is the house I live in, the village or town where I was born or where I spend most of my time. My home is my family, the worlds of my friends, the social and intellectual milieu in which I live, my profession, my company, my workplace. My home, obviously, is also the country I live in, the language I speak, and the intellectual and spiritual climate of my country expressed in the language spoken there ... My home, of course, is not only my Czechness, it is also my Czechoslovakness, which means my citizenship. Ultimately my home is Europe ... and – finally – it is this planet and its present civilization.

Havel (1991, 49)

The definition of home is an obvious starting place for a study of the meaning of home. Tuan (1971, 189) believes that "perhaps no single term in another language covers a significative field of comparable scope." Etymologically, the English word "home" can mean: "a dwelling place or house, a village or town, a collection of dwellings (Old and early Middle English); the place of one's nurturing, with the feelings which naturally and properly attach to it; a place, region, or state to which one properly belongs, in which one's affections centre, or where one finds refuge and rest" (Hayward 1975, 3). However, home in French, *maison*, refers to the physical structure, while the German *heim* connotes refuge or asylum. To these may be added the Old Nordic *heimr* for homeland and world; the Gothic *haims* and the Greek *kome* both translate as village.

Rybczynski (1986, 61) echoes this linguistic theme: "This wonderful word, 'home,' which connotes a physical 'place' but also has the more abstract sense of a 'state of being,' has no equivalent in the Latin or Slavic European languages." Sopher (1979, 130), analysing the meaning of "home," "neighbourhood," and "place," also examined these words in different languages. He provided a new perspective by suggesting that reference to home (town) and home (land) implies all of the warmth, security, and intimacy associated with references to home as a family dwelling.

To rely on one definition of the word "home" is misleading, and it is tempting to follow Kim Dovey's lead and suggest that "all of its uses in everyday life constitute its meaning" or that "home is a notion universal to our species, not as a place, house, or city, but as a principle for establishing a meaningful relationship with the environment" (Dovey 1978, 27). Box 1 presents a series of meanings of the word "home," and even this brief sample points to disparate meanings, often influenced by the perspective of the writer. The quotations in the following box are chosen for their attempt to provide a summary statement about "home." As such they suggest a common sense of refuge, possession, attachment, affection, and personal freedom.

Meaning of the Word "Home"

- "Home" is a label applied voluntarily and selectively to one or more environments to which a person feels some attachment (Hayward 1975, 3).
- Loewy and Snaith, following a study of consumers in the US housing market, reported the central concepts of home as:
  – a place to raise children/family
  – a place to live/stay/spend your time
  – a place to rest/relax/be comfortable in
  – a place for love/warmth/understanding
  – a place that I own/is my own/belongs to me
  – a place for privacy/to be alone/get away
  – a place you can always come home to
  – a place to be independent/can do as I please/security
  (Loewy and Snaith 1967, cited in Hayward 1975, 3)
- "Home" brought together meanings of house and household, of dwelling and refuge, of ownership and of affection. "Home" meant the house, but also everything that was in it and around it as well as

the people and the sense of satisfaction and contentment that all these conveyed. You could walk out of the house, but you always returned home (Rybczynski 1986, 61).

- Home is the place where one loves and is loved; it is a place where I go to rest in which I feel secure enough to lower my guard and lie down to sleep; home is where I keep my possessions; home is a place of comfort where pleasant experiences take place (Shaw 1990, 230).
- "house"/"home" (place to live in). The distinction was once more clear-cut than it now is. A "house" was a building for living in. A "home" was a "house" (or flat or family residence) seen as not just a place to live in but a place of domestic comfort and family happiness. Today the two words are – at any rate in the jargon of real estate agents – one and the same thing: "new show 'homes' for sale" ... In senses other than 'house,' however, 'home' remains a highly emotive word, as in 'homeland,' 'homesick,' 'home town,' and even the 'Home Guard' (Room 1985, 122).

Another approach to understanding the notion of home would be a brief description of the changing use of home as a physical structure or social concept in its European–American context. In addition, a review of trends in home decoration suggests that emphasis on comfort and particular styles reflects the importance placed by society on the creation of "hominess."

Homes, or, in this case, dwellings, were once more public; for example, the medieval lord's home had great halls full of servants and visitors, while the homes of artisans included their workshops and shops. In *Home*, a history of housing, technology, and social attitudes from the Middle Ages to modern times, Rybczynski (1986) traces the development of home as a concept. In the fourteenth century, townhouses combined living space in an upper area of one single large chamber and working space at the lower level. Beginning in the 1600s, homes, at least the more affluent ones, contained "privacies," rooms in which the individual could be sheltered from public view. In the seventeenth and eighteenth centuries, homes became the scene for domestic rather than working life and comfort gained new importance.

Rybczynski believes that the evolution of home comforts was gradual, encompassing the introduction of electricity, the disappearance of servants, and reappearance of the small family home. But the emphasis on home comforts became much more prevalent after the Exposition Internationale des Arts Décoratifs et Industriels Modernes held in

Paris in the summer of 1925. This extraordinary focus on the home interior featured pavilions highlighting the glamour of elaborate decoration and lighting and the *Esprit Nouveau* of the famous architect and designer Le Corbusier. The scene was set for modern home decoration.

While the emphasis on home decoration continued throughout the twentieth century, there have been changes in the general social attitudes toward home. In the 1950s, Jackson (1952, 6) wrote that "the modern American home, even the modern farm home, is fast becoming little more than a place where members of the family (not all of them, by any means) eat one or two meals a day, sleep, and enjoy occasional sociability." The late 1960s and 1970s in North America can be seen as a period when self-fulfillment meant more than attachment to anyone or anywhere, hippies and communes being the most obvious example of the footloose lifestyle. But more recently, the baby boomer generation has bought homes and there has emerged "a new awareness, a slowing down, a search for roots, family ties, a passion for the 'natural' and for the land, that powerful symbol of connectedness" (Johnson 1982, 9). This may be seen as a return to the pervading theme of home as a central cultural value and a means to stabilize society (Wright 1980, 294). The *Communitarian Manifesto* (Gwyn 1992, A5), published in November 1991, seeks "an active citizenry concerned about the moral direction of the community." Clinton's 1992 American presidential election campaign promoted "changing values" to strengthen the family and community. This emphasis has translated into concern about the family and social values of home. US sociologist Amitai Etzioni contrasts this movement with the environmental movement: "We have had, and still have, and still need, an environmental movement. What we need now is a social environment movement to heal society in the same way we're trying to heal nature" (Gwyn 1992, A5).

Reflecting these values, home decoration for the privileged has returned to more traditional themes. Designer Ralph Lauren mimics various historical periods in his home fashions and is "not so much interested in recalling the authentic appearance of a historical period as he is in evoking the atmosphere of traditional hominess and solid domesticity that is associated with the past ... a desire for custom and routine in a world characterized by constant change and innovation" (Rybczynski 1986, 9). The British designer Laura Ashley's "whole philosophy centred around the home, the family ... making products that make people feel comfortable, cosy" (Markoutsas 1992, C1).

Yet a third approach, but a formidable task and thus beyond the scope of this work, would be a review of people's own histories of home. The History Workshop movement has been responsible for the development of a "people's history," often created by ordinary people writing about themselves and frequently creating the only histories available that describe the lives of women and children. Biographies, while often telling of human relationships and social class, less frequently discuss the meaning of home to their subject (Porteous 1989, 232). However, together with fictional accounts, these sources would richly augment the study of the meaning of home. Similarly, the lives of previous inhabitants can sometimes be traced through their homes (see the box below).

Home as History of its Residents
- She is cordial as I leave, but she has told me she likes being alone. Of course she isn't alone at all. The place is filled with her predecessors (Johnson 1982, 112).
- To dwell means to inhabit the traces left by one's own living by which one always retraces the lives of one's ancestors (Illich 1985, 8).
- The corner to the right of the front door is the one that fifty years ago held an umbrella stand and where my father ... deposited a dripping wet umbrella; and where for twenty years hung a horseshoe found by my uncle Corrado (Levi 1989, 25).

The approach taken here, however, has been to examine the commonalities in the works of several major commentators who have studied the concept of home most extensively. Yi-Fu Tuan (1971) pointed out the complexity of the concept of home in terms of its etymological roots, the antinomic relationship of "home" and "journey," the sense of rest and the nostalgia associated with home. Hayward (1975) developed an overview of the multiple meanings of home while studying it as an environmental and social concept. Beginning with common dictionary meanings and then through readings in history, myth, and literature, he found descriptions of physical structures that are primary places of residence, descriptions of home as territory or a locus in space, and descriptions of home as a place of self-identity or as a social or cultural unit. In a later publication (1976), based on a study of a small sample of young residents of Manhattan, Hayward identified nine attributes of home:

- relationships with others
- relationship with community

- self-identity
- privacy and refuge
- relationship with other sources of meaning about home
- personalized place
- base of activity
- relationship with parents and place of upbringing
- relationship with a structure or shelter

Depres (1991) has reviewed Hayward's classifications and that of six others to create a revised typology. This definition of home involves:

- security and control
- reflection of one's ideas and values
- acting upon and modifying one's dwelling
- permanence and continuity
- relationship with family and friends
- centre of activities
- refuge from the outside world
- indicator of personal status
- material structure in a particular location
- place to own

Rakoff (1977, 93–4) interviewed a panel of white, middle-income people in Seattle and found the meanings of home to be a "multi-vocal symbol" that included: physical shelter, commodity or investment opportunity, place in which child rearing and family life occurs, indicator of personal status and success and sense of permanence and security.

Mackie (1981) defined two main themes of home within which sub-themes were discussed: (1) home as centre included the relations between home and identity, home and dreams, house and self, and home and away; and (2) home as refuge, covering protective qualities of the home and historical origins. Viewing home as a "principle for establishing a meaningful relationship with the environment," Dovey (1978) saw home as a place to which a person is attached, as security, as possessed territory, as the familiar, and as a base and starting place. In a more recent article, Dovey (1985) expanded these considerations to include among the properties of home the following: spatial order, temporal orientation, socio-cultural order, spatial identity, and temporal identity. He also identified the processes related to home that are expressed as: the spatial dialectics of home and journey; insideness and outsideness; order and chaos; and the social dialectics of self/other, identity/community, and private/public. Additionally, he contrasts bringing our meaning to our homes versus homes conferring identity upon us.

Tognoli (1987, 657) explored six aspects of home: centrality, rootedness, and place attachment; continuity, unity, and order; privacy, refuge, security, and ownership; self-identity and gender differences; social and family relationships; and socio-cultural context. Yet another theoretical framework for the meaning of home is proposed by Sixsmith (1986), who used a multiple sorting task plus in-depth interviews to determine the different meanings that home holds for people. She found twenty collective categories of which the six most frequently mentioned, listed in order of frequency, are: belonging, happiness, extent of services, self-expression, spatiality, and type of relationship. Watson and Austerberry (1986, 93–7) identified meanings of home to include decent material conditions and standards, emotional and physical well-being, loving and caring social relations, control and privacy, and simply a living/sleeping place. Somerville's (1992, 533) typology is based on a search for the meaning of home in order to define homelessness. He finds key signifiers for home including shelter, hearth, heart, privacy, roots, abode, and paradise and contrasts these with those of homelessness: lack of shelter, lack of hearth, heartlessness, lack of privacy, rootlessness, lack of a fixed abode and purgatory.

Additionally, Rapoport (1992) analysed the term "home" in both popular (folk) and professional use and recognized the following aspects of it: affective core, security, control, being at ease, relaxed, ownership, kinship, feeling comfortable, family, friendships, laughter, contentment, personalization, and taking possession. Finally, following interviews with people affected by a renewal project, Wikström concluded that home should not be interpreted as a scientific concept (1994, 318). He found home to mean subjective things such as warmth, comfort, and safety. Home was also a point of departure, a sense of autonomy, an opportunity to mutually create space, a place filled with memories, a sequence of events, and a part of a neighbourhood.

Together, these studies emphasize home and relationships – particularly family and friends – as well as the belief that home creates and supports identity, provides shelter, gives privacy and security, and is the predominant centre of our lives. A lesser theme is found in the combination of personal and material status and ownership as these concepts relate to home. The most recent works have focused upon home as the source of emotional well-being, comfort, and happiness.

Our content analysis of all these sources suggests that three major categories are salient in creating a typology of home, groupings that are more general and somewhat broader than the work of previous commentators. These aspects of home are: the spatial and physical; the symbolic meanings; and the psycho-social. Finally, we consider the meaning of home to an exile or homeless. This latter category is an initial attempt to reflect upon what it means to lose one's home, an issue to be taken up more thoroughly later.

HOME AS PLACE

I live in my house as I live inside my skin: I know more beautiful, more ample, more sturdy, and more picturesque skins: but it would seem to me unnatural to exchange them for mine.

Levi (1989, 25)

Home – Cluster of Meanings
- Home is the space/group/time entity in which individuals spend the greater part of their lives. It is preferred space, and it provides a fixed point of reference around which the individual may personally structure his or her spatial reality (Porteous 1976, 390).
- The concept of home is applicable across all scales from the individual psyche, the room, the house, the street, the neighbourhood, the town to the nation and the globe. Home can refer to a physical entity such as a cave, a house, an orphanage. On an experiential level, home can refer to the daily round of life in one's habitual abode (Mackie 1981, 2).

Spatial aspects of home are expressed as a cluster of meanings, as illustrated in the box below. From this, several themes emerge: home as a hierarchy of physical places, the dichotomy between private and public space, home as the core node or centre of one's activity space, and the physical appearance of home. Generally, and despite the American attempt to define home by what we carry around in our car (Appleyard 1979, 18), the spatial concept of home is conceived as a series of concentric zones ranging from one's own room, to one's dwelling, neighbourhood, village, town, or city; region, nation or country; and finally the whole world. Each of these levels of home can be considered as a separate focus of attachment, with the levels of dwelling and surrounding neighbourhood or landscape being the most relevant to the concept of domicide. But whole regions can also be affected:

Room: "My home is the room I live in for a time, the room I've grown accustomed to, and which, in a manner of speaking, I have covered with my own invisible lining" (Havel 1991, 49).

Dwelling: "As a home, the house is a creation having special properties accessible only to the people who made it their home. These properties ... are difficult to portray from the outside. This is because in its deepest sense, home is always something personal and private" (Karjalainen 1993, 70).

Neighbourhood: "To be forcibly evicted from one's home and neighbourhood is to be stripped of a sheathing, which in its familiarity, protects human beings from the bewilderment of the outside world" (Tuan 1974, 99).

Village: "The villager who has never moved away ... retains the unique mark of his particular village. If a man says that he comes from Akenfield, he knows that he is telling someone from another part of the neighbourhood a good deal more than this. Anything from his appearance to his politics could be involved" (Blythe 1969, 18).

Landscape: "Pioneer records are rich in examples of settlers forming unusually strong attachments to the familiar features in the landscape" (Rees 1982, 1).

Region: "The commonest core lies in a widespread feeling of belonging someplace, of being 'at home' in a region that extends out from but well beyond the dwelling unit" (Fried and Gleicher 1961, quoted in Hayward 1975, 6).

Nation: "O Canada! Our home and native land!" (Lavallée and Routhier 1880).

Earth: "To be at home on the planet and welcome here, humanity must understand and appreciate the primacy of that home, the Eden we have never left, and the wild that is its emblem" (Rowe 1990, 34).

The concept of home as a hierarchy of places may also be seen as a clustering at various spatial levels; for example, there is a link between room and dwelling. The dwelling is then set in a neighbourhood and the neighbourhood in a village, town, or city (or in the case of a rural area, in a landscape). Finally, all of the above are found within region or nation. Research in France suggests that humans are most attached to the levels of the dwelling and the nation (Burgel 1992, 4). Thus, at the subnational level, the dwelling and its immediate surroundings are the chief focus of an individual's spatial concept of home. This level, which may include one's neighbourhood, appears most relevant to the concept of domicide to all except the world's remaining nomads, modern-day gypsies, and New Age travellers whose lifestyle is ostensibly free from such absolute constraints.

Within those spaces that are recognized as home, private, semi-private, and public spaces are recognized. Bollnow viewed the house as the means by which "man carves out of the universal space a special and to some extent private space and thus separates inner space from outer space" (Bollnow 1960, 33). Greenbie (1981) sees the significance of the single family dwelling as a transition in space between our own bodies and the outside world, a transition that is assisted by the provision of windows, fences, and thresholds. This dichotomy of private versus public space and the intervening thresholds (porches, steps, front yards, backyards, driveways, sidewalks, and alleys) is also explored by Taylor and Brower (1985), who conclude that these spaces, emanating from the home, help to define the behaviour of the immediate community.

The threshold of a home has particular significance because it is the division between public place and private sanctity, and because thresholds vary depending on the cultural norm. While Americans may have open unfenced front yards, the English often have a fenced front garden with a gate. Moslems, on the other hand, have high walls around their compounds (Rapoport 1969). Cultural differences in the need for privacy are expressed in this way.

Altman and Gauvain (1981) also discussed the role of thresholds as well as that of windows in terms of the accessibility that these parts of the home provide to residents. After studying victims of burglaries, Korosec-Serfaty (1984) found that the boundaries between the inside and the outside of the house are essential features of dwelling experience, the door providing the boundary between the outside and the inner self. These commentators recognize the important transition that occurs between the inside and the outside of homes, between private and public spaces, and they emphasize the significance of the interior of the home – the most intimate element of home.

Viewing home in the context of its larger setting, it can also be described as a core node within a nexus of nodes which comprises the individual's activity space (Porteous 1977, 93) (see the box below). Home is the place from which one starts out and to which one returns after a day's work – "a still point in an ever turning world" – an irreplaceable centre of human significance and existence (Hayward 1975, 6; Relph 1976, 39–40). This concept is of fundamental importance to any study of domicide, for if home is the "centre of the world," then losing home is "undoing the meaning of the world" (Berger 1984, 56–7).

Discussions of home, the physical dwelling, often focus on outside appearance in which both public/community and private/individual values are reflected. At the community level, dwellings reflect certain values through the use of materials and design, while at the individual

---

Home as Core Node/Centre

- Nothing can be *done* without a previous orientation, and any orientation implies acquiring a fixed point. For this reason, religious man has always sought to fix his abode at 'the centre of the world' (Eliade 1957, 22).
- And lastly, in the name of fire, which controlled, is the greatest friend of man and uncontrolled, his most relentless enemy; greatest of forces; worshipped since the most ancient times; focusing point of mankind. The family gathers about the fireplace; the Indian lights his tepee fire; and where the pioneer far from civilization makes his tiny blaze, that spot is home (W.D. Richardson in Engel 1983, viii).
- The creation of a centre of sanctity in a profane world is the beginning of order in space. This centre is the germ of the home. The centre of consciousness that is self is realized in the environment as the centre of radiating binary pairs. Many characteristics of home previously described radiate from this centre – familiarity in a strange world, security within insecurity, certainty within doubt, sanctity within profaneness, order within chaos, passive sanctuary in an active world (Dovey 1978, 28).

---

level, specific details that reflect the inhabitants and their societal ties become more evident, particularly within the home (Altman and Chemers 1980; Altman and Gauvain 1981; Bothwell, Gindroz, and Lang 1998). This situation is perhaps exemplified by Santa Fe, New Mexico, where adobe is used as the building material throughout but where the design details such as gates or courtyards of individual buildings create a constantly changing image.

Werner et al. (1989, 280) find a relationship between home appearance, personalization and upkeep and ethnic identity, social class, lifestyle preferences, and religious identification. By choosing to live in a home having a certain external appearance, a person may also be expressing how they wish to be seen; for example, certain housing developments appear very ostentatious (Porteous 1976, 384), while others communicate attitudes such as attachment, openness and neighbourhood sociability. Werner et al. (1989) found, for example, that strangers identified friendly home exteriors by the presence of Christmas decorations.

When we are invited inside, we experience the wonderful variability of home: the decoration, atmosphere, and meaning of its rooms. Weisner and Weibel (1981) studied interior home environments and found four major distinguishing characteristics: disorder/functional complexity, decorative complexity, warmth/child-orientedness, and the presence of books. Further, they found that values and cultural/lifestyle choices rather than material conditions were the strongest predictors of home-environment differences. This thought is echoed by Johnson (1982, 5): "How they spoke, all these rooms, but how often they communicated things their owners never intended." Exploring the more difficult concept of atmosphere within the home, Pennartz (1986) analysed the experience of pleasantness in rooms and found a correlation with various spatial characteristics such as their size and shape. Some rooms are found to be special places for the assertion of self-identity. In a study by Korosec-Serfaty (1984, 303), the attic and cellar were often identified as the secret spaces of the home. She found that these spaces may have negative connotations for some, while for others, they signify shelters, allowing accumulation and security.

Traditional societies enclosed sacred space within their homes, which, like the hearth, helped to "unify natural, social, and supernatural realms and to resolve symbolically the conflicts among them" (Rakoff 1977, 86). Pastoral nomads orient temporary shelters and body positions in relation to their fire as the centre of their geographic area (Dovey 1978, 28). For example, both Mongolian nomads and scholars refer to the mandala, "The Yurt and the Universe," which shows the brazier at the centre, surrounded by the hearth square, and then the domed skin tent, the yurt. This is bounded conceptually by a box representing the four corners of the Earth and a circle representing the Earth (Faegre 1979, 93).

Home is not only the centre of our world but centre *and* whole. While today needlework decorations of "Home Sweet Home" or "God Bless This Home" over the hearth are fairly rare, the family still centres around the electronic hearth or the dining table. Since home is the centre of life for most human groups, domicide involves loss of this centre.

HOME AS SYMBOL

For most people, there is a transformation of the experience of space or a piece of land into a culturally meaningful and shared symbol, that is, place. The symbol (place) then evokes the transformed experience and reminds us of its cultural meanings and social implications.

Low (1992, 286)

Just as place transforms to a symbol, so does home in most of its manifestations. Westerners place individual, psychological symbolism on their homes; for example, the Navajo Indians attach cultural or

collective symbolism to their round dwellings. In contrast, the formerly nomadic Basarwa of the Kalahari Desert in Botswana attach no symbolism to their dwellings (Kent 1992, 3), but this is rare. The most
common array of symbolic meanings of home includes: home as a
memory of past experience, home as a source of nostalgia, and idealized or imagined conceptions of home. Home also carries an ideological sense in terms of homeland or private property. For some people,
home may also mean the grave or God.

Memories of home frequently pervade our reminiscences about the
past or are revealed through psychoanalytic means. Tuan (1971, 190)
believes that the word "home" is more applicable to an accumulation
of past experiences than to the immediate reality of home. But home
is also a "memory machine," causing us to relive our past experiences
through its contents (Douglas 1991, 294). Attempting to separate out
our memories of home in order to extract the essence of dwelling
places, Bachelard (1964, 29) concluded that home brings memories
and dreams together: "the house protects the dreamer [and] is one of
the greatest powers of integration for the thoughts, memories, and
dreams of mankind." This intertwining of memory and imagination is
also recognized by Tindall (1991, 221). She reminds us that in Jung's
dream of a house, there were different floors sheltering different activities, and in the basement were old bones, memories hidden beneath
our consciousness. The theme of remembered home and all the
warmth and affection centred therein is further illustrated in the following box.

Home as Memory or Memorial
- Typically, the home is set in the past, in memories of childhood, as a
  "*recherche*" for the "*temps perdu*," the home of memory, which is the
  only basis for a sense of identity which the exiled writer can maintain
  (Gurr 1981,11).
- If the meaning of home lies in the accumulated memory of each past
  day, it also lies in an expectation of future days. In our memory, home
  is peaceful, reassuring, and comforting but as we experience each day,
  home also holds darkness and sadness (Mackie 1981, 43-45).
- Memories of the outside world will never have the same tonality as
  those of home and, by recalling these memories, we add to our store
  of dreams; we are never real historians but always near poets, and our
  emotion is perhaps nothing but an expression of a poetry that was lost
  (Bachelard 1964, 29).

- The accumulation of consecutive rooms in his memory now resembled those displays of grouped elbow chairs on show, and beds, and lamps, and inglenooks, which, ignoring all space–time distinctions, commingle in the soft light of a furniture store beyond which it snows, and the dusk deepens, and nobody really loves anybody (Nabokov in Tindall 1991, 221).
- We are all profoundly affected by the places we live, often without realizing it. Their problems and paradoxes become our own, changing us and making us part of them. When we leave, the memories of their rooms and streets stay in our minds like ghosts, or the voices of old lovers (Johnson 1982, 6).
- Their little tract of land serves as a kind of permanent rallying point for their domestic feelings, as a tablet upon which they are written, which makes them objects of memory in a thousand instances when they would otherwise be forgotten. It is a fountain fitted to the nature of social man from which supplies of affection, as pure as his heart was intended for, are daily drawn (Wordsworth cited in Lucas 1988, 87).

Where home provides a lasting memory in the form of a memorial, a strong link between home and identity is found. In exploring the meaning behind landscapes, Lucas (1988, 89) discusses Wordsworth's poem *Michael*. He suggests that the destruction of the cottage named The Evening Star and the land surrounding it strewn with stones means an end to memories: "[the stones] tell of broken hopes, of the destruction of continuity, of the obliteration of a family and even of community, for they had been gathered for a sheepfold which Michael had intended to build with and for his son, who was to have been the inheritor of his land." The land and buildings are thus marked with a human significance that outweighs any value they may have as picturesque objects. Similarly, for the poet Clare, "to change the look of the land was to wound the lives of those who lived on and through it ... altering the landscape obliterates 'objects of memory'" (Lucas, 89).

Memories of home often result in intense attacks of nostalgia. According to Tuan (1971, 189), the word "nostalgia," from the Greek *nosos* (return to native land) and *algos* (pain or grief), was coined in 1688 by Johannes Hofer, a medical student who believed that this "homesickness" deserved medical attention. Since the concept of home is now writ large across popular magazines (for example, see "Thoughts of Home" in the American publication *House Beautiful*), it seems just possible that home is currently vying with nature as the post-Romantic or postmodern replacement for God. Nostalgia is indeed rampant (see the following box).

Home and Nostalgia

- We have more than the ever-persistent nostalgia ... for some simple and quiet home where we can recapture long-lost values swept away by social change (Johnson 1982, 7–9).
- They were returning to very diversely imagined paradises, [but] nostalgia was common to them all (Holt 1966, 131).
- Be it ever so humble, there's no place like home (Payne 1823).
- Home again, home again, jiggety jig (from nursery rhyme *To Market, To Market*)
- You can't go home again ... and for starters, let us admit right off that Thomas Wolfe hit the Motherlode back in 1934. But wouldn't Tom have gone out of his tree had he had any notion of what that piece of information would mean today; 'cause, why, "You Can't Go Home Again" has been squared and cubed and raised to the fourth power! (Anonymous 1972, 131)
- I want to look for Ne-Hi Pop and Burma Shave signs and go to a ball game and sit at a marble-topped soda fountain and drive through the kind of small towns that Deanna Durbin and Mickey Rooney used to live in in the movies. It's time to go home (Bryson 1988, 43).
- I thought: as soon as all this is finished, we'll go straight back to Brixton. ...We could have stayed out there, cut ourselves a nice little niche out there, hit the high spots. But I was pining away for home – for the whirr and rattle of the trams, the lights of Electric Avenue glowing like bad fish through the good old London fog, longing for rain and weather and bacon sandwiches, for the healthy chill of 49 Bard Road on a frosty morning, for the smell of home, the damp, the cabbage, the tea, the gin (Carter 1982, 142).
- Should we not think of the problem of home and peace ... as only a bit of nostalgia or a remnant of whispering from the past but finally melting away in the age of postmodernism? (Karjalainen 1993, 72).
- It is not our exalted feelings, it is our sentiment that makes the necessary home (Bowen 1986, 140).

It is a common experience that a certain slanting of the light, or the smell of a spice from our past will bring home flooding back to us (Norris 1990, 239, Porteous 1990). Proust's remembrance of his childhood in Combray on tasting and smelling *petites madeleines* is the most famous literary example of such an experience. The past, even the form of a remembered landscape, is essential in order to understand what we are seeing: "Patterns in the landscape make sense to us because we share a history with them" (Lowenthal 1975, 5). Buttimer recognizes that nostalgia for place, and particularly for rural settings, is often experienced by persons enveloped by urban surroundings. She suggests that such feelings are strongest during periods of significant change in either social or physical environment (Buttimer 1980, 166). Indeed, social mobility in many countries has ensured that nostalgia for home has become endemic (Hardyment 1990, 12).

Nostalgia frequently involves idealization. The specification of an ideal home will, of course, vary depending on the individual. Thus, Tuan contrasts the Alaskans' liking for their "frozen landscapes" with the Nuer's for the swamps of the Sudan (Tuan 1974, 114). Homes can also become the embodiment of fantasies (Johnson 1982, 4), the manifestation of an ideal that is realistic or not and, as such, provide a place of escape (see the following box). Felicité in *La Fortune des Rougon* sits in a window gazing at the Place de la Sous-Préfecture, which was a "small square, bare, neat, with nice light houses, [and] seemed Eden to her" (Zola cited in Tindall 1991, 47).

Ideal/Images of Home

- Though I live by choice in the city, home is a rambling country house in some place where there is snow on the ground. There are fireplaces and many bookcases and deep carpeting everywhere. Though I can scarcely sew on a button, my dream home is a showplace of handicrafts, all created by me (Johnson 1982, 2).
- We invest in places where we live with a lifetime of images about home (Johnson 1982, 7).
- People have intellectual, imaginary, and symbolic conceptions of place as well as personal and social associations (Buttimer 1980, 166).
- Home may also be idealized, a preconceived image of what home should be (Cooper, 1974)
- Such idealized images give a false notion to reality of everyday experience of home. The experience of home is not bound to any one ideal form but is as variable and as valid as there are individual life trajectories (Mackie 1981, 28–9).
- It grows in the sun and sleeps in the stillness of night; and it is not dreamless. Does not your house dream? And dreaming, leave the city for grove or hilltop (Gibran 1965, 31).

Homes sometimes take shape as a product of our imaginations. While the widely-fantasized luxury of building one's own home is less frequent today, such a home must surely be the ultimate example of our imaginings. Rybczynski (1989, 191) describes Carl Jung's country retreat, which was created over a period of thirty-two years in a style intended to emulate the past. It was, in Jung's words, "a kind of representation in stone of my innermost thoughts." Idealized visions of home, frequently seen as a detached house in a rectangular yard, may provide a false sense of reality, for home can vary from apartment to park bench (Cramer 1960, 41; Porteous 1977, 65).

Beyond imagined or idealized visions of home, Bachelard (1964) has influenced a whole generation of writers with his concept that the house protects the dreamer. Like Rybczynski (1989, 190), many may have first thought this an "obscure conceit," but upon reflection, they found it quite reasonable, for home is where "it is safe to let our minds drift." And dreams of home are not only of the built environment but also of a favourite place where you go to dream, where "topophilia" is manifest (Tuan 1961). Such dreams may soon extend to cyberspace in which virtual reality permits the "morphing" of impossible dreams and, for the artist: "Going home, feeling home, [will not be] as easy as it once was. We've still got the instinct, but someone has thrown away the map" (Creighton-Kelly 1992).

A more down-to-earth example of the idealized home is found in the acquisition of a home in a suburban area. While the earliest suburbs housed poorer segments of the population "outside the city gates," the richer elements of society also found a place for larger mansions or summer homes. By the late 1700s, suburbs in England had taken on a new air of respectability and, later still, improvements to transportation heightened the trend for suburban living. Blythe (1981, 22) acknowledged how suburbs in the twentieth century became "part of a comfortable and preferred way of life of half of Britain's population." The suburb is also home to half of the American population and to much of the population in Australia. As such, it presents a mixed picture as an ideal. Comment about and criticism of suburbs, urban sprawl, and the attendant problems abound and have spilled over into fiction, from Orwell's *Coming Up for Air* to Bowen's *Attractive New Homes* (Tindall 1991, 237). Nevertheless, Blythe believes that the general preference for this type of home should not be ignored.

Older suburbs have come to resemble inner cities with problems of seniors and the homeless and with a decaying infrastructure, but new suburban developments continually attempt to recreate new sorts of idealized homes. Bearing names like Heartland or Green Valley, some recently constructed suburbs have been described as "a seamless facade of interminable, well-manicured developments punctuated by golf courses and an occasional shopping plaza done in stucco" (Guterson 1992, 55). The developments Guterson describes, which have sprung up in the desert outside of Las Vegas, are not, however, without problems of crime (such as drug abuse, and gang violence) as well as the intrusion of emissions from nearby heavy industry.

While achieving an ideal home may be impossible, understanding the image of home that people hold becomes crucial in planning new homes. The advantage of suburban-type homes may have been seen as their provision of a "clearly evidenced universe" in their opportunity for "authentic living" as defined by Bachelard who proposed that the complete experience of home includes homes with cellars and attics, snuggling into villages, or rising in the middle of fields (Marc 1972, 137; Korosec-Serfaty 1984, 305).

Ideals may, of course, become ideology. The ideological sense of home is expressed in terms of home as a right, the sanctity of private property, and the concept of homeland involving patriotism. Berger (1990, 85) suggests that the word "home" has been taken over by two kinds of moralists: the defenders of domestic morality and property (including women), and those who defend the notion of homeland.

Hollander (1991, 31) illustrates home as a right with reference to Robert Frost's *The Death of the Hired Man*:

> 'Home is the place where, when you have to go there,
> They have to take you in.'
> 'I should have called it
> Something you somehow haven't to deserve.'

Hollander also suggests that, while home may not be an earned right, it may become so when home is one's private property. When first legally possessing a home, it does not immediately feel like your own, for it takes some time to erase the signs of ownership of previous residents (Lang 1985, 202). Even when a home becomes one's own property, there are still limits on its use, usually expressed in the form of building codes and zoning bylaws. These limits have been enshrined through numerous legal precedents including the common law principle of "nuisance." On what is perhaps a lighter note, the US Supreme Court held that a man's home is indeed his castle, but his backyard is just the castle's "curtilage" and enjoys much less privacy, particularly when that curtilage is used for growing marijuana (Will 1986).

While the importance of home ownership may vary between countries and cultures (for example, it is less important in Switzerland than it is in Australia, Canada, and the United States), Rakoff (1977, 94) found that his sample of white, middle-income people in the Seattle area continually returned to the assumption that ownership was necessary for such things as permanence, security, control, status, refuge, and family life. Rakoff also suggested that "for most people, home ownership, current or imagined, is the single most important characteristic of the house, in large part because ownership helps them to resolve the conflicts and ambiguities that the private, home space is heir to" (Rakoff 1977, 100). Authors specializing in this subject area (Denman 1978, 2; Macpherson 1978, 179; Ryan 1987, 72) have provided broad descriptions of the concept of property as a social and jurisdictional institution; a vehicle of power in human relationships; a determinant of the occupation, possession, and ownership of land; a provider of opportunities for freedom, self-expression and allegiance; and the boundary between the individual and the state. Like "home," "property is *not* an object but rather a social relation that defines the property holder with respect to something of value (the benefit stream) against all others" (Bromley 1991, 2).

At some stage, homes often change from private property (the ultimate bastion) to a product (the ultimate exchange commodity). Rakoff (1977, 88) sees the house as a crucial commodity of the political economy as well as the scene of much of everyday life. This contradiction is recognized by Heidegger (cited in Relph 1976, 40), who feels that home is a perverted phenomenon when expressed in terms of monetary value, and also by Raskin, who notes that, while "the idea of 'Home' tends to be relegated to sentimental songs and sayings ... the actuality is a series of residences built, sold, and occupied as generally replaceable commodities" (cited in Hayward 1975, 4). Hardyment (1990, 12) also explores this theme and, in considering the effect that moving to a new house may have on children, decries the way in which homes are considered to be simply a financial investment. She says: "Isn't it a form of prostitution, this decking of houses in seemly shades of Dulux and knicker-blinds from Laura Ashley, and putting them up for sale as desirable residences?"

In an ideological sense, home may also mean homeland (see the following box). Exploring this concept, Schama (1991) contrasts the wartime visions of homeland by the British artist Frank Newbould, which showed a stone-walled village nestling at the base of undulating hills, with a German poster by Bergmann showing a ploughman with a strong horse tilling fertile acres. Two quite different versions of home are portrayed by these artists; they wished to take advantage of the notion that landscape and people are one, therefore relating (and perhaps manipulating) their countries' image of what is important in wartime. Even today, the themes of art often focus on ideals such as home, place, nationalism, refuge, and safe spaces (Creighton-Kelly 1992). A particularly strong recognition of a sense of home is also found in accounts of pioneers who chose a particular place to settle because it reminded them of their homeland (Rees 1982, 3).

Home/Homeland

- The claim that landscape and people are morphologically akin, constructed, as it were, from common clay, and that they constitute in some primal cultural sense the nature of each other – that land and homeland may be interchangeable – is now a familiar commonplace (Mack 1991, 11).
- At a larger scale, America is experienced as home (Sopher 1979, 129).
- Human groups nearly everywhere tend to regard their own homeland as the centre of the world (Tuan 1977, 136).
- Marta tells me of the violation of her house, of the door kicked in by soldiers. And yet, although her house was destroyed, its very shelteringness desecrated, she is quick to add that Guatemala, without politics, is paradise. How can this be – that we yearn for hell and call it paradise? (Norris 1990, 238).
- 'Tis the star spangled banner, O, long may it wave
  O'er the land of the free, and the home of the brave.
  (Key 1814).

A final ideological sense of the word "home" is its use to mean the grave, heaven or God, although this is now less commonly used (see the following box). Ecclesiastes 12.5 identified home as heaven and the place of ultimate return (Hollander 1991, 33). Gurr (1981, 13) suggests that Donne equated his need for God with his need for home. In keeping with the previous discussion of home as ideal, Gurr believes that today "the ideal of God as our home has tailed off into a pallid cliché, an unconvincing assertion of wish fulfilment, the idea of getting away from it all taken as far as it will go." It is, perhaps, this loss of the notion of heaven as providing the ultimate home which makes clinging to the earthly home more poignant.

Home Meaning the Grave/ Heaven/God
- ... and the grasshopper shall be a burden, and desire shall fail: because man goeth to his long home, and the mourners go about the streets ... (Ecclesiastes 12.5)
- For though through many straits, and lands I roam,
  I launch at paradise, and I sail toward home (Donne).
- "Turn up the lights," he protested to the nurse, adding in a paraphrase of the popular song of 1907, "I don't want to go home in the dark." (Langford 1957, 245).
- If there was one of those tiny graveyards behind the house with its cluster of family graves, my envy became awe. It was a sacred place as well, their home. They had buried their ancestors on their land ... home, according to Reynolds Price, is a religious place containing our dead (Johnson 1982, 103).
- O God, our help in ages past, ... and our eternal home (Watts 1930, Hymn 662).
- No, I cannot believe that. I hold another creed, which no one ever taught me, and which I seldom mention; but in which I delight, and to which I cling, for it extends hope to all. It makes Eternity a rest – a mighty home, not a terror and an abyss (Brontë 1971, 51).

HOME'S PSYCHOSOCIAL MEANING

Man is born homeless; and the search for home
Creates him and destroys him hour by hour.
　　Herbert Reed cited in Tindall (1991, 213)

The psychological and social aspects of home are explored now in terms of the meaning of home at various stages in the life cycle: by role or relationship (spouse, parent), by feelings toward home, and by the relationship of place to self. In this context, home is where the heart is, "an ideological construct created from people's emotionally charged experiences of where they happen to live" (Gurney cited in Sommerville 1992, 529).

Traditionally, people were born at home and might also die there. Indeed, the periods of childhood and old age predominate in discussions of the psychosocial meaning of home (see the following box). For the child, home provides the centre, the mould, the place where socialization and acculturation occur (Appleyard 1979, 6; Hobsbawm 1991,

66; Porteous 1990, 157). Home is the place where the child first "learns to understand his being-in-the-world" (Norberg-Schulz in Relph 1976, 42); later, it is where a person's "looking-glass self" learns to interpret how others react to home and thus to themselves (Gunn n.d., 18). When students at the College of Environmental Design at Berkeley were asked to design an ideal living environment, they frequently included aspects of their childhood surroundings (Cooper 1974).

Stage in Life
Child
- Sarah and I often spent cold, rainy afternoons playing "In My Mansion" to escape our less than satisfactory childhoods (Johnson 1982, 1).
- It is from the home that we begin our journey into the world beyond the immediate space of the house that we live in (Winning 1990, 246).
- In his text "The Homecomer," Schutz (1971) describes home as "starting point as well as terminus." By this, I believe he means that our journeys, in a broad sense of the word, begin and end in a specific place (Shaw 1990, 227).

Old Age
- I've seen it here, elderly black people don't fit in nursing homes here, they get bitter and disillusioned, very miserable if they can't return. When you're old, you should be among your own. You can't let go of your roots, they're something to hold on to. With retirement and illness, it's better to be there (Western 1992, 22).
- As one grows older, a man returns to his roots, so closing the circle of his life (Vassilikos 1991, 310).
- It is the actual geography of boyhood and girlhood which the old long for (Blythe 1980, 41).
- It is certainly emotionally consoling, to both parent and child, to consider that the parental household is, in many ways, a home permanently available to those who have been nurtured in it. "This will always be your home; it is here whenever you need or want to come here" is the traditional speech, so grating to whatever family the child has acquired on his own (Martin 1984, 244).

Home is the centre that provides initial protection, and from which forays for exploration and escape are made. Through direct observation of children in a New England town and through in-depth interviews, Hart (1979) observed this ever-widening range from home. Home may be a place of refuge for the child, but when home is less than satisfactory, an imaginary home may be created as a place of escape (Johnson 1982, 1). The environmental psychologist Mary Janskoski believes that our childhood homes put a permanent imprint on our neurological abilities: "You think it only translates into preferences, but it actually affects our nerves" (Kyriakos 1994, D9). Perhaps it is here that the formation of the idealized home begins.

Children's books often seem to involve a quest and return to home, with a small "action space" to permit the return to home as quickly as possible. In fact, for many of us, "home may well really mean our childhood home" (Porteous 1990, 143). Home is also more central to adults when children are at home (Appleyard 1979, 18), when, as parents, they provide the centre of security for their children.

Chawla's (1994, 1995) work on environmental autobiography demonstrates how important their childhood memories of home may be to adults, but it also finds that women have a far greater and usually more positive attachment to their childhood memories than do men. Says Mrs. Boyle in Urquhart's *The Underpainter* (1997, 101): "I'm very fond of places ... I always pray for my three most special places ... the farm in Kerry where I grew up, the farm we had to leave to come here, and the little house we have now. I pray for them every night."

In old age, there is the desire, even when far from home, to return to one's home of origin (Western 1992). Both Rowles (1978) and Blythe (1980), in their interviews with the elderly, recognize the significance of childhood landscapes. For some, however, last days are spent in euphemistic "homes": homes with special names like old folks' homes, nursing homes, retirement homes, sunset homes, and mental homes. "In the psychiatric wing, no one speaks of home," and people lose their identity (Porteous 1990, 186). Porteous (1976, 388) also emphasizes the one-way nature of the journey to the old people's home and the consequent decline in health. This is similar to the effects caused by relocation during urban renewal. For example, in Lieberman's (1983) study of 639 elderly people, half were either dead, physically impaired, or had deteriorated psychologically one year after they had changed their living arrangements. Those near death were withdrawn and passive, which was the only way they were able to cope with their circumstances. To make matters worse, some elderly people are moved several times as they come to need different levels of care. How different it would be if the elderly were permitted to participate in the design of their future homes (Boschetti 1990) choosing the colour of walls or where to place furniture or personal possessions.

In the more central stages of the life cycle, feelings toward home vary according to the role of individuals and their relationships to others. Home is often inseparable from family, and it may be defined in terms of the responsibilities of the husband or as the workplace for wives or for women in general (see the following box). An examination of the meaning of home to women, as seen through "people's history," is particularly interesting because, traditionally, women have spent more time at home than men.

Home/Role or Relationship

- There is an infinite difference between the home we choose and the one that is chosen for us. Women follow men to the places where they work, and it falls to them to carve out a hollow in the new space and make it work, and men, who choose the places, more often remain apart from the life of the place where they live (Johnson 1982, 10–11).
- For me, home is inseparable from family. It is within the fold of the familiar that I am allowed to be me. It is where I don't need to explain who I am (Norris 1990, 242).
- "Home" is some place where I could be happy, whether it's here or Barbados doesn't matter. Where it's happy, warm, you can have friends come into, can do what you want, can have a laugh and a joke, and play a few records or something. In many ways, this is more my home, ours, we've worked, saved, got this place, made it what we want it to be. The Barbados one was my *parents'* "home" (Western 1992).

Rybczynski (1986, 160) suggests that feminization of the home began in the seventeenth century in Holland and reached its peak in nineteenth-century America in the work of Catherine Beecher and Harriet Beecher Stowe. In this work, it was recognized that "woman's place was in the home [but] that the home was not a particularly well-thought-out place for her to be." For some, this feminization of home has negative connotations. McDowell and Massey (1984, 128–30) compare the lifestyles of women in nineteenth-century County Durham and modern Hackney. In the former, the women's role of unpaid work in the home, in which "working clothes had to be boiled in coppers over the fire which had to heat all the hot water for washing clothes, people, and floors," supported the filthy and dangerous business of mining. Modern women in Hackney provide another definition of the "woman's workhouse" as they are involved in piecework for the rag trade in their own homes: "I used to get my work done in five hours,

now I work ten or twelve hours a day ... The kids say, mum, I don't
know why you sit there all those hours. I tell them, I don't do it for
love, I've got to feed and clothe us. I won't work Sundays though. I
have to think about the noise ... I'm cooped up in a cupboard all day
– I keep my machine in the storage cupboard, its about three feet
square with no windows." For these women from different eras, the
sense of confinement and entrapment is the same.

For others, particularly American middle-class women prior to
1940, home was viewed in a more positive light as the setting for
domestic arts. The transformation of house to home was a traditional
role in which "the atmosphere of home was seen as having an almost
mystical effect on its inhabitants, determining their moral standards,
happiness and success in the outside world" (Motz and Browne 1988).
Such an attitude to home is typified by the following:

> A homekeeper am I: this is my task
> To make one little spot all snug and warm,
> Where those so bruised and beaten by the day
> May find refuge from the night and storm.
>
> Gladly I serve – love makes the serving sweet;
> I feel no load – love makes the burden light;
> A happy keeper I of home and hearts –
> Serving I reign – a queen by love's own right.
> F.J. Hadley, *The Mother's Magazine*
> (cited in Davison 1980, 67).

In a discursive journey through women's magazines, Davison
(1980), traces three generations of American women and their homes
and finds that women are now developing a more ambivalent attitude
to the pleasantries of domestication. However, this domesticity of
home may not be such a bad thing, for as Hardyment (1990, 12) points
out, "home as nest" may not be as smart as a "des. res. with all mod.
cons.," but it may well be preferable.

Two studies have confirmed that this feminization of home is still
prevalent despite the growing number of women working outside of
the home. Ahrentzen et al. (1989) studied the use of space in homes by
538 family households in Toronto. They found that fully employed
married women spend more time in rooms with family members and
are more involved in housekeeping and child-care activities than their
spouses. Tognoli (1980) also found that more women than men were
involved with activities in the home. However, this circumstance may
be changing. Mui (1992) studied the arrangement and allocation of

space within a house and finds that as societal attitudes and the status of women and men change, the use of space within the home also changes. For example, women often now require "a room of one's own" despite their continuing responsibilities for the care of all the home.

Other studies have provided information about the physical arrangements of home and how they reveal the relationships of those who live there. Peled and Ayalon (1988) describe a therapy in which a couple were made aware of how the meanings they invested in the spatial organization of their ideal homes revealed the conflict in their relationship. Similarly, Irwin (1992) studied polygamous Mormon families in Utah and found that environmental setting played an important role both in the viability of relationships between the husband and each of his wives and the relationships between the wives. As a result of these analyses, more therapeutic spatial layouts were produced that assisted in their therapies.

There is often a distinction between one's own home or the family home as a social unit and the home of parents, particularly when the latter is in another country (Western 1992). The parental home is frequently the site of traditions, and it can be oppressive to the individual (Appleyard 1979, 4; Porteous 1976, 387). But new homes are established and new traditions begin. When ten women and men who lived alone were asked to describe their experiences after leaving their parental homes, they outlined three phases: an initial phase of feeling "not at home;" then an awareness of a need for home; and, finally, the psychological and physical arrival at a place that felt like home (Horowitz and Tognoli 1982). The authors interpret these findings to suggest that home can have various environmental and psychological dimensions over time and that the meaning of home does not necessarily depend on traditional family structures.

Feelings toward home are most frequently expressed as affection (see the following box). Feelings of "at homeness," "about home," and "being at home" form the connection or measure of quality between person and home. We often choose to make homes the settings for important rituals such as birthdays, weddings, or funerals (Saile 1985). Metaphorically, we speak about being at home with people or with an idea when we are comfortable with them.

The geographic literature about home is broadly contained within this continuum that links place, home, and feelings about home, the latter connecting people to place. Relph (1976, 1) asserts that "to be human is to live in a world that is filled with significant places: to be human is to have and to know *your* place." Hay (1987, 1) describes sense of place as resulting from both residence (in one place) and

Feelings Toward Home
- Home is where the heart is (attributed to Pliny, quoted in Hayward 1975, 2).
- Where we love is home, home that our feet may leave, but not our hearts (O.W. Holmes, quoted in Hayward 1975, 2).
- Attachment to place is defined as "the symbolic relationship formed by people giving culturally shared emotional/affective meanings to a particular space or piece of land that provides the basis for individual's and group's understanding and relationship to the environment" (Low 1992, 286).
- Our house was not insentient matter – it had a heart and a soul, and eyes to see with; and approvals and solicitudes and deep sympathies; it was of us, and we were in its confidence and lived in its grace and in the peace of its benedictions. We never came home from an absence that its face did not light up and speak out in eloquent welcome – and we could not enter it unmoved (Mark Twain cited in Rybczynski 1989, 171).
- For a man who prefers not to be happy, there is the attic where he can listen till nightfall to the creakings and groanings of shipwrecks, or the open road where the wind will blow his scarf against his mouth like a sudden kiss that brings tears to the eyes. But for one who loves happiness there is a house by the side of a muddy footpath, the house at La Sablonnières, the door of which has just closed on my friend Meaulnes and Yvonne de Galais, who became his wife at noon (Alain-Fournier 1966, 161).

awareness (of that place). He traced the study of sense of place through various areas of geography. Much of the literature involving sense of place is relevant to a discussion of the meaning of home as it links human behaviour and habitat. However, that discussion is limited here in order to focus on literature directly related to home. Beyond feelings of affection for home, this literature has expressed other concepts including psychological territoriality, rootedness, security, irrational attachment, and refuge.

Hayward (1975, 5–6) developed clusters of meanings about home. One cluster included home as territory, tying the physical space of home and neighbourhood (home area, home range, hometown) to feelings of familiarity, belongingness, and predictability, and acting as a spatial framework for behaviour. Porteous (1976, 384) discussed the ethological concept of territoriality in relation to home and found that home provides the individual and the family with a triad of satisfactions: identity, security, and stimulation. The territorial imperative is very strong where home is concerned. As Porteous suggests: "the average citizen appears to expend more effort personalizing and defending the home than any other level of fixed physical space."

Feelings toward home may involve a bonding that is so close as to be described as rootedness or significant attachment. This sense of rootedness may be the test for being authentically at home (Relph 1976, 41), or at least it "implies being at home in an unselfconscious way ... [where] human personality merges with its milieu" (Tuan 1980, 4). "To be rooted is perhaps the most important and least recognized need of the human soul" (Weil 1952, 43). Feelings about home may also involve an almost irrational attachment, whereby people are so attached to their homes that they are satisfied to remain within them, virtual hermits, contenting themselves with diversions that come to them (Johnson 1982, 145). This attachment is also found where people have worked the land and built their homes. Writing about the poem *Michael*, Lucas (1988, 87) provides a new construction on Wordsworth's words:

What makes the landscape where Michael lives cherishable has nothing to do with its picturesque properties, as "you" are brought to realize ... it has to do with endeavour, work, and all that is contained in the key terms: "occupation," "abode," "dwelling." It is because of these things that the fields and hills where Michael dwells:

> ... had laid
> Strong hold on his affections, were to him
> A pleasurable feeling of blind love,
> The pleasure which there is in life itself.

Home also provides a sense of security or refuge, feelings that relate even to the most rudimentary form of shelter (see the following box). This pleasure in the rituals of home – the importance of prized objects and collections and their care, the sense of embeddedness and bonding – is brought out in a number of studies (Csikszentmihalyi and Rochberg-Halton 1981, Boschetti 1995). Cooper-Marcus (1995) sees home and its objects as a mirror of the self. Little wonder that the elderly hang on to their prized possessions. When his mother cleans out his dead grandfather's apartment and gets rid of all his possessions, it is to the grandchild in Bygrave's (1974, 107) novel "as if his grandfather had never lived." Those who are not attached to their homes discuss them in a manner that differs significantly from those who are attached (Twigger-Ross and Uzell 1996).

Home as Refuge/Security

- Dunroamin came into its own when vast numbers of upper-working- and lower-middle-class families from dreary Victorian terraces and rural slums just managed to scrape their way into a situation which, slump and war threats notwithstanding, gave them the feeling of secure anchorage (Blythe 1981, 22).
- Home is one sure refuge for persons forced to frequently venture beyond it. The "house as haven" is not a lifestyle confined to the lower class (Porteous 1976, 386).
- Home is for me as yet a fortress from which to essay raid and foray, an embattled position behind whose walls one may return to lick new wounds and plan fresh journeys to further horizons (Maxwell 1963, 210).
- And now the house is immune from the outside world, of which the only reminder is the scratching sound made against pane by the leafless branch of a rose bush. Like passengers on a boat adrift, the lovers, in the wintry gale, are enclosed in their own happiness (Alain-Fournier 1966, 167).
- For almost everyone, the notion of home is usually a positive one. It is the known as opposed to the unknown; it is certainty as opposed to uncertainty, security rather than insecurity, the knowledge that in the final analysis, someone else, our parents, will make the necessary decisions and will protect us from harm ... even where homes are inadequate, children may choose to stay because it is the only real experience of home and parent that they know. It is familiar and predictable (Shaw 1990, 227).
- Home may become a defensive symbol protecting the family against "the spectre of destitution," and the bewildering outside world (Appleyard 1979, 5).
- This is the true nature of home – the place of Peace; the shelter, not only from injury, but from all terror, doubt, and division (Ruskin cited in Welsh 1971).

Western homes provide a physical defence against threats, with parlours for meeting guests and windows for surveillance (Gunn n.d., 2). Who will enter and what will take place inside can be controlled (Dovey 1978, 27). The significance of such control is seen in modern day home-security arrangements. The cartoon character Calvin ponders as he wakes to a slight 2:00 a.m. sound: "When someone breaks into your home, it shatters your sense of security. If you're not safe in your own home, you're not safe anywhere." And then: "A man's home is his castle. But it shouldn't have to be a fortress" (Waterson 1989, 10).

Several commentators have stressed the sense of security provided by home. At home, you can "leave the world and be alone, safe from danger"(Marc 1977, 14). This is seen as particularly true for low-income groups for whom home becomes haven (Gans 1967, 27; Willmott and Young 1957, 269). By contrast, Vidich and Bensman (1960, 58) note that for professionals and skilled workers, home is a different form of security – an accumulation of equity. More significantly for this study, Bollnow proposes an "anthropological function of the house" in which the feeling of security provided is essential for self-identification (Egenter 1992, 6). If home means security, it is all the more devastating when home is invaded or destroyed.

Security may also lie in a strong sense of self-identity. Within the literature of home, there are frequent discussions of the meaning of home to self – the way in which home shapes identity or in which homes are shaped by the inhabitant (see the following box). This relationship is argued in differing ways. Self is seen as the most important, among other considerations, in the integration with home; being at home is defined as being close to self. Home is a second body, which is seen as a symbol of self and self-identity. Home shapes you and, in turn, is shaped in your image. Home may change you against your will or without your knowledge. Ironically, the strong sense of self created by a strong sense of home may also be the factor that preserves you when home is lost.

Place-identity/self-identity
- Don't I, when I go into other people's homes, draw the most sweeping conclusions about them ... aren't they offering it up for view, like a road map of their very souls? (Johnson 1982, 2).
- No act of self-scrutiny can be complete if we don't see ourselves in relation to these outer shells and understand their influence on us ... I have lived for varying periods in five widely disparate places since I was a child. Because of the nature of each place, and because of what I was when I lived in them, each changed me – usually against my will, for I fought them all, probably for the same reason we fight intimacy, the fearful yielding that might hurt (Johnson 1982, 12–13).

- It's not even enough difference to make me a mongrel. I never really noticed the difference until I went back there, again, and found how different from home my home was. You don't choose your own landscapes. They choose you (Carter 1982, 24–5).
- For even as you have homecomings in your twilight, so has the wanderer in you, the ever distant and alone. Your house is your larger body (Gibran 1965, 31).
- In the life of a man, the house thrusts aside contingencies, its counsels of continuity are unceasing. Without it, man would be a dispersed being. It maintains him through the storms of the heavens and through those of life. It is body and soul (Bachelard 1964, 29).
- To be at home, to dwell authentically, is to be incorporated into the landscape of home, which is the place from which we have our sense of who we are (Winning 1990, 257).
- While the motives for choosing a home are still instrumental (sufficient space, good access, and various creature comforts), economic, and aesthetic, the symbolic role of the home as an expression or confirmation of desired identity is increasingly important, though seldom discussed except in critical terms (Appleyard 1979, 4).

Zonn (1983, 3) explores the place-identity/self-identity perspective of home. For him, place-identity is the unique character of a place as seen and interpreted by the individual, while self-identity is the way persons see themselves as unique from other persons. When place-identity and self-identity are closely allied, a strong attachment may result, but this does not necessarily result in a home. Zonn argues that home involves the intersection of self, centre, rest, authenticity, and insideness. Similarly, Dovey (1978, 28) suggests that to be "at home" or to "make oneself at home" is to act naturally and come closer to one's self. Confirming the special, close relationship between home and self, Lang (1985, 202) sees home as "the intimate hollow we have carved out of the anonymous, the alien" – in fact our second body. The analogy of the house as body is echoed by Bachelard and Gibran (see the preceding box). Based on studying literature, poetry, dreams, and contemporary architecture, Cooper (1974) found the home to be an important symbol of self and self-identity.

Studying urban identities, Proshansky et al. (1983, 57–9) theorized that there are a series of functions that connect place identity and a sense of belonging. These functions seem to represent the various stages in the link between home and identity, with the final stage representing the most territorial aspect. The five stages comprise the following:

- *recognition* function, which provides an environmental past against which the immediate physical setting can be measured
- *meaning* function, which suggests what should happen in places and what is appropriate behaviour
- *expressive requirement* function, which is the expression of tastes and preferences regarding the space
- *mediating change* function, in which change in a space is required
- *anxiety defence* function, which occurs when other people must be involved in any change to a place

Appleyard (1979, 5) has also explored a number of ways in which home and identity are linked. These include: the creation of our identities and traditions during the time we are resident in our first family homes, the special bond that is created when people have the privilege of building or designing their own home, and the way in which choice of home may express identity. However, Appleyard cautions that any discussion of the way in which home reflects identity must recognize that "We can be, successively, the person we would like to be, the person we wish we were not, and the person we think we know we are." This caution is echoed by Proshansky et al. (1983, 80), who note that little attention has been paid to sex, class, ethnic, and other group differences in considering the interactions between the person and the physical setting. Finally, the identity created by home can also be limiting. Concerned that designers who are captives of past residential situations could not develop innovative housing or housing for different socio-economic or cultural groups, Ladd (1977) encouraged architecture and planning students to reconstruct their residential histories and thus break down these constraints.

Augmenting these academic sources are discussions of the relationship between place and identity or self in what can only be a minimal selection from poetry and prose. The intertwining of identity and home in literature has roots at least as far back as the poet Horace, but it reached its apogee in eighteenth-century verse and nineteenth-century novel. Cowper, Crabbe (*The Village*), Goldsmith (*The Deserted Village*), and Clare exemplify the former. In Clare's "The Flitting" and "To a Fallen Elm," the subject is the interconnectedness of place and identity and the loss that results from separating person from place (Lucas 1988, 89); in the case of "The Flitting," it is the loss of Clare's "home of homes" and the familiarity of his neighbourhood. From a wide array of possibilities, George Eliot's *The Mill on the Floss*, Emily

Brontë's *Wuthering Heights*, and Charlotte Brontë's *Jane Eyre* provide good examples of the latter. Rootedness and identity are the essence of Mr. Tulliver's relationship with the mill; neither the dead Cathy nor the living Joseph can easily relinquish Wuthering Heights; Mr. Rochester suspects Jane Eyre is becoming attached to his house through the operation of "the organ of Adhesiveness."

The relationship continues in twentieth-century literature. E.M. Forster's Mrs. Wilcox identifies passionately with *Howards End*, while Margaret Schlegel detests a "civilization of luggage" and expects to end her life "caring most for a place." In *Seasons of My Life*, the story of Hannah Hauxwell's lone struggle to survive on a desolate farm in the Yorkshire Dales, the theme of place-loyalty and identity is exemplified. She says of her farm: " Wherever I go ... and whatever I am ... this is me."

HOME TO THE EXILE OR HOMELESS

Just so, the migrant's adopted home is never home, but the migrant is too changed to be welcome in her own country. Only in dreams will she see the skies of home. The ache of exile cannot be assuaged by travelling anywhere, least of all by retracing old steps looking for houses that have been bulldozed and landscapes that have disappeared under urban sprawl and motorway.

Greer (1993)

Home and journey have been offered as the fundamental dialectic of human life (Tuan 1971; Porteous 1976), but for Porteous (1990, 107), this dialectic is home and away. The ultimate expression of this antinomy may be home and non-home: exile or homelessness. There is a major cultural assumption that home is not only who you are but where you come from – your cultural milieu (Creighton-Kelly 1992, 18). Life away from home is considered inauthentic, despite the place of exile being the new home. Winning (1991, 180) drew upon her experience of teaching English as a second language to formulate her ideas regarding home and away: "At home, people speak to each other in a particular way; At home, there is more laughter; An accent comes from somewhere else; When away from home, we hear the sound of words; *and* The talk of home is different."

The literature relating to exile (Simpson 1995) is sufficiently large to be the subject of a separate book if studied from the perspective of travel writing, fiction, and poetry; only a brief glimpse is offered here in order to elicit the feelings about the meaning of home of those exiled (see the following box). The feelings of some of these prodigals for whom exile is self-imposed, for whom elsewhere is better, is not denied.

Home – Exile and Homelessness

- There was probably no place on earth that can have seemed less like home, no atoll or hill station or desert oasis that can have been less sympathetic to the peculiar needs of the wandering Englishman and his family. But, like the good colonist that he was, he did eventually manage to fashion the place into an approximation of Surrey-in-the-Sea (Winchester 1985, 119).
- But the British who were out in Burma were not engaged in new ideas, new books, or new ways of being men and women ... They imported Life as They thought It Was, with confidence (Wiggins 1989, 25–7).
- The symbolic character of the notion of "home" is emotionally evocative and hard to describe ... home means one thing to the man who has never left it, another thing to the man who lives far from it, and still another to him who returns ... [it] is an expression of the highest degree of familiarity and intimacy (Schutz 1971,107–8).
- The house had been lived in by strangers for a long time. I had not thought it would hurt me to see it in other hands, but it did. I wanted to tell them to trim their hedges, to repaint the window frames, to pay heed to repairs. I had feared and fought the old man, yet he proclaimed himself in my veins. But it was their house now, whoever they were, not ours, not mine (Laurence 1989, 191).
- Meanwhile, the wafts from his old home pleaded, whispered, conjured, and finally claimed him imperiously (Grahame 1908, 86).
- Children in India are greatly loved and indulged, and we never felt that we were foreigners, not India's own; we felt at home, safely held in her large warm embrace, content as we never were to be content in our own country (Godden 1966, 9).
- This sense of home is the goal of all voyages of self-discovery which have become the characteristic shape of modern literature. In varying degree, the normal role of the modern creative writer is to be an exile. He is the lone traveller in the countries of the mind, always threatened by hostile natives (Gurr 1981, 13).

For others, the longing for home while far away is quite intense. Such feelings might well be similar to those of a person permanently exiled from home by domicide.

It is suggested that writers begin with a sense of home as base and source of identity. In exile, the writer's loss of home provides a sense of perspective and a point of comparison but may also anticipate for others a loss of identity, history, and sense of home (Gurr 1981, 14). Porteous (1990, 141), in a discussion of the work of Graham Greene, suggests that Greene continuously points out the "inauthenticity of existence away from home," even though he found it necessary to escape home. In fact, Greene entitled his autobiography *Ways of Escape*. Gerard Manley Hopkins, who believed that he had been exiled from family, friends, and homeland by prejudice against his religion, constantly sought home in new places (Martin 1991). Jack Kerouac's whole life on the road was a search for home, and his writing recorded that yearning (Nicosia 1990, 19).

For exiles, home may be their country of origin, a former residence, or both. Western (1992) interviewed thirty-four expatriates from Barbados living in London. When asked, "When you use the word home, what are you thinking about?" the largest single response was "Barbados"; the second, "It depends upon the context"; and the third, "I've got two homes, Britain and Barbados."

Exiles who suffer homesickness often attempt to make their new place of residence as much like home as possible. Schutz points out that to live in a land of strangers is to live in a place in which their past and ours do not cross, as if we have no history (cited in Norris 1990, 240). The modern writer in exile, particularly in Britain and the United States, often seeks home through the recreation of a past time and a focus on cultural heritage (Buttimer 1980, 166; Gurr 1981, 14). The desire to mimic homes of origin seems to have been particularly prevalent among colonial exiles (Winchester 1985, 119; Wiggins 1989, 25–7) and permitted the desired escape from the reality of the outpost. According to Gurr (1981, 15), James Joyce, in exile, spent his literary life trying to recreate his Dublin home in minute detail. In India, hill stations such as Simla and Ootacamund mimicked the distant English way of life. For Canadian pioneers, the feeling of homesickness frequently passed when a place was found where the landscape looked like home (Rees 1982, 1). In order to recreate home, Emily Carr's father cultivated an English garden in the nineteenth-century British Columbia landscape (Carr 1942).

Returning home also has special meaning to an exile or frequent traveller. Shaw (1990) speaks of the elaborate plans for a return home, the unfulfilled expectations when things are not as they once were, and the pleasure of a return to security. Some exiles, of course, are unable to return home because their home has been destroyed; the home that they have marked as authentic, that has formed a point of comparison, and that has been the focus of memory is lost forever.

But homes may be lost without the trouble of exile, and homelessness is becoming more prevalent. The United Nations International

Year of Shelter for the Homeless (1987) broadly defined the homeless to include those who have no homes, such as "street people" and disaster victims, and those who are relatively homeless in terms of their living standards (Charette 1987, 4). Writers on homelessness frequently define the subject on the basis of lack of control, inadequate privacy, and poor material conditions but neglect more emotional issues such as indifference, powerlessness, and anomie (Sommerville 1992, 530-33). Feelings about the absence of home create a counterpoint to expressions regarding the positive qualities of home.

However, there is also another side to any discussion of homelessness, which relates to the feelings held by the homeless about the substandard conditions in which they have to live. For many people, squatting allows them to have some sort of home when they would otherwise be completely homeless. Their own home may no longer exist for them because of eviction from their own or family land; they may be unable to stay in that home or on their land due to economic circumstances; or they may have left due to social or political pressures. In general, the desire for a decent home, or any kind of home for that matter, is behind squatting, and examples of it occur throughout the world. As early as 1980, one-tenth of the global population were classified as squatters, although there were many different names for it: shantytowns in North America, *bidonvilles* in France, *gourbevilles* in Tunisia, *favelas* in Brazil, and *colonias paracaidistas* in Mexico (Gimson 1980, 206). In 1987 in Papua New Guinea, squatters formed as much as one-third to one-half of the population of the larger towns (Mason and Hereniko 1987, 141,173). There is a form of squatting that uses existing but empty houses, and over a quarter of a million people in Britain took over housing that did not belong to them between 1960 and 1980. As one individual declared: "I Peter Manzoni, restorer...having noticed that the premises known as 29 Winchester Road were open, unoccupied, and in an advanced stage of decay, entered thereon with the express intention of *creating a home*" (Ingham 1980, 166).

Homelessness is another subject that deserves a separate study. It is simply observed here that, like exile, homelessness speaks of the loss of identity experienced by those away from home and that squatter communities demonstrate the desperate desire to regain some semblance of home. When the former home has been destroyed, however, this is clearly impossible. Yet there have been few attempts to study the destruction of home and place, including the motives for, and process of, destruction and the human reactions to the loss of home and place. The present study will attempt to remedy this deficiency.

HOME AND DOMICIDE

Home is a complex, multi-faceted phenomenon, a term voluntarily applied to places of varying size. Home has been found to have several meanings that can coexist, without contradiction, within an individual. Each person's experience relating to such an intimate subject as "home" appears to significantly colour the interpretations of home's meaning: "home was something complicated, 'irrational,' and, at the same time, very important" (Wikström 1994, 317). For example, women have traditionally experienced a more intense process of home-making, cultural differences may affect the expression of feelings about home, home means different things to people of different ages or in different roles, and home may be important to a person only when it is private property. Most important is the close relationship between person and home, the emotional bonds that exist. More than anything else, home is where the heart is, and while feelings and attachment to home vary between individuals, all of these factors, when combined, provide a rich description of the meaning of home.

Both humanistic-literary and empirical-behavioural research demonstrate that all aspects of home interpenetrate, and that home has complex, multiple, but interrelated meanings. Werner et al. (1985) believe that home can be seen as only a "holistic transactional unity" in which separate aspects (physical, symbolic, socio-psychological) are considered in an integrated manner.

Two focuses appear to perform this integrative task. The first, an outward-lookingness, focuses upon *home as centre* – a place of refuge, freedom, possession, shelter, and security. The second, an inward-lookingness, focuses upon *home as identity* – with themes of family, friends and community, attachment, rootedness, memory, and nostalgia. Home is thus a spatial, psychosocial centre in which at least a portion of an individual's or a group's identity resides.

Having reached an acceptable, if contestable, definition of the essence of home, it remains to consider those aspects of home that are of significance when home is lost. To fully understand the chapters that follow, it is important to summarize those aspects of home of particular interest in any consideration of the effect of domicide. These factors are found within the general groupings of spatial, symbolic, and psychosocial meanings of home.

Within the category of spatial meaning, it is found that home is primarily the dwelling (shelter) we live in. But our sense of home is also enriched by garden (or rural setting); neighbourhood; village, town, or city; and country or nation. All these are called home, and there appears to be no value to more narrowly defining the concept of home in its physical sense – certainly all can be subject to domicide. While Burgel's (1992, 4) research suggests that attachment is chiefly to dwelling and nation, most of the literature portrays a primary attachment to dwelling, the place where one lives. At the level of dwelling, home is a special place – inner space as opposed to outer space (Bollnow 1960, 31) – with definite connections to the outer world through thresholds, doors, and windows. Within the home, placement of furniture and decoration express comfort and identity, particularly if the residents have lived there for a long time, say for more than one generation. Domicide, even when a new home is found, means that home is never the same. Change is inevitable. Home has the sense of being permanent – the core, the centre from which other characteristics of home emerge, the home of identity, the place to return to. Destruction of it implies "undoing the meaning of [this] world" (Berger 1984, 56–7). Together, homes define a community and its values. When whole communities are permanently destroyed, quite obviously the loss is all the more significant.

The symbolic meaning of home is epitomized by Hannah Hauxwell as she gazes at the home she must leave (and which through her good fortune will not be destroyed) and says: "they cannot take it away from me, it's mine, mine for the taking and always will be ... even when I'm no longer there" (Hauxwell and Cockcroft 1989). Home is the "memory machine" (Douglas 1991, 294), and while memories may be good or bad, in memories of home as with childbirth, the pain may be put aside. Home is the place in which memories and dreams meet, and identity is formed. Domicide erases the physical place of memory and source of identity not by conscious choice, as when one changes homes, but through the deliberate acts of others. It is suggested that the loss is worst for those whose home has become a living memorial – when they worked on the land or the structure of the home, and when it is meant to be passed on to future generations.

Home in the context of nostalgia was discussed by Tuan (1971, 189). Since 1971, commentators on the subject of home have given this concept less prominence, yet any review of current popular literature will suggest that nostalgia for home is a predominant theme. Nostalgia does not appear to be "melting away in this age of postmodernism" (Karjalainen 1993, 72) despite the effect of shrinking physical distances due to advances in transportation and communication technology. But the postmodern period still retains the desire for economic expansion, and, thus, homes and their settings are continually destroyed. For some, nostalgia for home may be all that survives.

Home is portrayed in ideal, imagined ways, and for those whose home is destroyed, this may be the only hope to which they can cling.

The possibility of finding and creating their ideal home is all that comforts them as they are forced to move on. Home is also considered a basic right, and the concept of private property is sacrosanct, particularly in North America and Europe. It is this factor that will set a significant measuring point in requesting compensation for those who lose their homes through domicide.

Psychosocial aspects of home have been defined in terms of the meaning of home based on stages in the life cycle, role or relationship, feelings toward home, and by the relationship of home to self. Home is where the heart is, "the locking together of self and artifact" (Bordessa 1989, 34). Home is significant at all times in our lives, but particularly when we are children. Home is the centre, the mould, to family and friends; it is our place in the community and the place where acculturation occurs. Feelings toward home include psychological territoriality, rootedness, a sense of security, irrational attachment, and refuge. Given these feelings, the invasion and/or destruction of home is likely to be all the more devastating. As Bowen (1986) has written: "After internal upheavals, it is important to fix on imperturbable *things*. These things are what we mean when we speak of civilization: they remind us how exceedingly seldom the unseemly or unforeseeable rears its head. In this sense, the destruction of buildings and furniture is more palpably dreadful to the spirit than the destruction of human life." The link between home and identity is the subject of a large number of studies reviewed above and is common in literary sources. The sense of identity created by a sense of home may therefore be destroyed when home is destroyed.

Discussion of the meaning of home to the exile or the homeless presents a similar situation to that which arises when home is destroyed. Predominant for the exile is the desire to seek or recreate home in new places. Yet the exile may eventually be able to return home, while others cannot because their homes no longer exist. The fact that their "object of memory" has been obliterated forever is one of the central problems of domicide.

Domicide, then, may result in: the destruction of a place of attachment and refuge; loss of security and ownership; restrictions on freedom; partial loss of identity; and a radical de-centring from place, family, and community. There may be a loss of historical connection; a weakening of roots; and partial erasure of the sources of memory, dreams, nostalgia, and ideals. If home has multiple, complex meanings that are interwoven, then so does domicide. All the complexity that defines the nature of home, when combined with the examples of domicide that will be explored in the next three chapters, support the belief that what is lost is not only the physical place, but the emotional essence of home – aspects of personal self-identity.

CHAPTER THREE

Extreme Domicide:
Landscapes of Violence

So I returned and considered all the oppressions
that are done under the sun:
and behold the tears of such as were oppressed,
and they had no comforter;
and on the side of the oppressors there was power;
but they had no comforter.
Wherefore I praised the dead which were already dead
more than the living which are yet alive.
Yea, better is he than both they,
who hath not seen the evil work
that is done under the sun.   Ecclesiastes 4: 1-3

Extreme domicide involves major, planned operations that occur rather sporadically in time but often affect large areas and change the lives of considerable numbers of people. Such events are not everyday occurrences for most of us and are often regarded in personal life histories or in collective memory as epoch-making episodes, which are organized not by us but by others for our benefit or detriment. Extreme domicide, then, will be considered below in terms of war, colonial geopiracy, and resettlement projects. These are not watertight compartments; there is considerable overlap. But running through almost all categories of extreme domicide is a marked strain of physical violence, reduced in only some cases to the level of mere coercion.

WAR

Cities, armies, agriculture:
Humankind becoming its own vulture.
J. Douglas Porteous (1997, 33)

"War is hell," declared General William Tecumseh Sherman, but national leaders, military careerists, and major business corporations often do very well as a result of it. Warfare has been with us since the beginning of Western civilization in Mesopotamia about 10,000 years ago when standing armies emerged along with cities, agriculture, literacy, organized religion, and social class hierarchies. Toward the end of the twentieth century, the world was spending almost $2 million per minute on its military forces (Thomas 1995, 132), and about 85 per cent of all war casualties were civilian. We are all targets now.

Goya's arresting painting *Tres de Mayo* appals us because we so readily perceive ourselves as the victims. Those about to be executed have faces, and the faceless soldiers are just carrying out orders. This is a universal idea; it is not just Madrid 1808, but Santiago de Chile 1973, Beijing 1989, and Grozny 2000. But we too often forget that killing people is not the whole objective of war. For war also kills places – witness Picasso's *Guernica* – and without loved places (homes, communities, landscapes, nations), a homeless person and faceless people easily lose identity and *raison d'être*.

The destruction of homes, places, and even homelands is an inevitable result of modern warfare. But it is domicide only if home destruction is deliberately planned. This is often the case, and home annihilation in war is pursued because of revenge, leverage against enemy governments, and in an attempt to terrorize civilians so that they lose faith in their own cause. Even public-interest rhetoric may be used; it is difficult to forget the American army's rationale for the annihilation of a Vietnamese settlement: "We had to destroy the town to save it" (Eliot et al. 1993, 542).

Domicide in war is discussed in terms of: scorched earth (the deliberate destruction of the physical, economic, and environmental bases of localities), strategic bombing, strategic resettlement, and the creation of military installations.

Scorched Earth

Modern total war, with its destruction of homes was prefigured by the early military operations of the Mesopotamian city states of the third millennium BC. Warfare in this era involved the complete destruction of small cities, the demolition of their temples, the carrying off of their inhabitants into captivity, and attempts at both memoricide and ecocide. The cultured antiquary and bibliophile Assurbanipal, for example, reported, with relish: "I levelled the city and its houses ... I consumed them with fire ... after I had destroyed Babylon ... and massacred its population, I tore up its soil and cast it into the Euphrates" (Ragon 1981, 276). In 146 BC, a similar fate befell Carthage at the hands of the Romans. Having razed the city to the ground, the Romans formally cursed it and sowed the site with salt. Little moved or grew there for over one hundred years. And after the destruction of Jerusalem by Titus in 70 AD, that city sank to the level of an anonymous Roman garrison town. Two thousand years later, the systematic devastation of enemy territory had become a commonplace. "War is war," platitudinized General Sherman as he pursued his scorched earth march across Georgia in 1864 (Dyer 1985, 79).

It was in Vietnam, however, that the scorched earth policy reached its zenith. The United States' attack on Vietnam was the first time that military technology had been employed in an attempt to destroy the environment of an entire nation. By the late 1960s, herbicides were being dropped on nearly one million acres of forest and cropland each year (Thomas 1995, 112). As if this were not enough, numerous "coconut raids" wiped out whole villages, along with their animals, fruit and coconut groves, paddy fields, and dykes. Twenty-ton bulldozers – known as "Rome plows," an echo of Carthage – were employed to systematically wreck villages and plow them under. Elsewhere, coastal villages had their mangrove swamps drained and burned. Cambodia and Laos fared no better. In Laos, the district of Xepon, close to the Vietnamese border, was one of the most heavily bombed. As one survivor recounts: "In our district there are two hundred villages. The bombing destroyed 6,557 houses – every house in the district. Five thousand buildings that were used to store rice were also destroyed" (Sesser 1994, 79). Here, then, destruction reached an intensity difficult for outsiders to imagine.

The Soviet Union committed similar atrocities in Afghanistan in the 1980s, while the 1990s began with the United States turning Iraqi towns into rubble in Operation Desert Storm. Saddam Hussein likewise mounted similar operations against his many minority peoples. In northern Iraq, many Kurdish villages were sprayed with poison gas and then bulldozed, as in the case of Halabja in 1988 (Hitchens 1992). After Desert Storm, the Western allies simply stood by as Saddam Hussein destroyed the whole environment of the "Marsh Arabs," who had lived on floating islands in the southern Iraqi swamps since Sumerian times. Other domicidal atrocities committed against minorities by Saddam Hussein include actions against Shi'ite Muslims, Assyrian Christians, Turkomans, and even Sunni opponents. Between the spring of 1987 and February 1988, the Iraqi government destroyed thirty-one Assyrian villages, including twenty-five monasteries and churches. In 1991, Saddam crushed a post-Gulf War revolt by destroying much of two Shi'ite holy cities. "Najaf and Kerbala were sacked as no Iraqi city has been sacked since the Mongols took Baghdad in 1258" (Makiya 1994, 23, 219).

Also seeking to create an indelible object lesson for his dissident countrymen, Syrian President Hafiz al-Assad destroyed much of the Syrian city of Hama in February 1982. Between 10,000 and 40,000 people were killed over a two-month period. Even after the clearing up process the returned lived on in fear: "You don't take photos, you don't do anything, you don't even look at the destruction, because if you stand there staring, they might arrest you. People in Hama were basically paralysed, in a state of shock ... My first shock was the people, people more than buildings because, you see, I was expecting to see physical destruction ... The next shock was the buildings ... those beautiful old buildings that I have such strong memories of ... People didn't identify with their town anymore" (Makiya 1994, 301–3). The returned exile goes on to describe his shock in viewing a bald bare hill, where once had stood "a magnificent panorama of old, closely packed homes ... Kelenyia [had been] surrounded by tanks and pounded into oblivion. Now they had tidied it up into this earth mound, a bald hill out of which there rose a tower built over the bones of all those dead people who are still there in the foundations. The tower is a brand-new Meridian hotel."

Strategic Bombing

The development of weaponry has increased the distance between the killer and victim exponentially. Rather than twisting his bayonet in the guts of his enemy, the modern warrior can annihilate thousands by bombing from 30,000 feet or even from another continent by remote control. Although modern land warfare can easily reduce villages and small towns to rubble, aerial warfare readily extends this destructive power to cities and regions. By the 1960s, a single B52 carrying a twenty-ton bomb load could destroy an entire village in a single blow, and a formation of B52s could obliterate a "box" over half a mile wide by four miles long. It is commonly known that the United States dropped three times the tonnage of bombs on Vietnam, Laos, and Cambodia as were used in the whole of the Second World War. Moving up the intensity ladder, the same country dropped twice as many bombs on Iraq in 1991 in a mere six weeks as were used in the entire Second World War (Thomas 1995, 122).

The strategy known as carpet, area, saturation, or pattern bombing, however, was brought to fruition during the Second World War. The deliberate devastation of cities and their residential areas was seen as a quick road to victory, costing money rather than allied lives. Based on studies of German bombings of Hull, Coventry, and other British cities, Lord Chartwell declared dispassionately in 1942: "Investigation seems to show that having one's house demolished is most damaging to morale. People seem to mind it more than having their friends or even relatives killed" (Dyer 1985, 91).

The Canadian geographer Hewitt (1983b, 1987, 1993, 1994a, 1994b) has made some interesting analyses of the effects of Second World War bombing on Britain, Germany, and Japan. He concludes the following: over 90 per cent of air-raid deaths were civilian; those bombed out and forced to evacuate totalled 1.4 million in London, 1.9 million in Berlin, and 2.9 million in Tokyo; in Germany alone, 3.3 million houses were destroyed; and the greatest losses occurred in the inner, more congested areas of cities. In fact, these raids were known unofficially as "slum raids," their targets being the homes of working-class women and children, the elderly, and the infirm. Significantly, not only were inner-city houses destroyed, but also many buildings of historic importance – ancient landmarks that were the symbols of identity and of the continuity of urban life and culture. Returning to mere ruins, an evacuee from Hanover remarked: "I have the feeling that I am a stranger in my hometown" (Hewitt 1994b, 20).

That the raids were vindictive there is no doubt. They caused the greatest distress to those very people who were least able to influence their home government's policies. Two quotations garnered by Hewitt (1994b, 2) demonstrate that the victims readily understood this point: "I thought the objective of aerial bombardment was to destroy the enemy's war plants and shipping, and not hospitals and the homes of the people" (male factory worker, Nagoya); and "Why did they bomb us – women, children, residential sections? Our men were gone. They were on the Front. Actually it was senseless ... There were no industries here" (woman, Wellingsbüttel, Hamburg).

The senselessness advocated by Air Marshall Arthur "Bomber" Harris and his ilk reached its climax in the raid on Dresden in 1945. Known as "the Florence on the Elbe," Dresden was a Baroque architectural marvel, a symphony of gracious streetscapes, with no significant war industries. Yet on the night of Ash Wednesday in 1945, an armada of 800 bombers caused a series of infernos, which in turn generated a firestorm. As the flames consumed the city's oxygen, air from outside was sucked in with the force of a hurricane, spreading the fire and suffocating those who were not burned to ashes. Such was the devastation that parts of Dresden's centre still lay waste in the 1990s. In 1995, embarrassed by their neo-Assyrian vengeance, the British offered to pay for the gilded cross that was to top the reconstruction of the destroyed Frauenkirche, once a Baroque masterpiece (Steele 1995).

The Americans targeted Tokyo in a similar way. Soon after Dresden, General Curtis LeMay ordered the first low-level mass night raid on the city, concentrating on an area 6.5 km by 5, with 30,000 inhabitants per square km. Over one-quarter of Tokyo's buildings were destroyed in this one raid, and 1.008 million people were made homeless. This was followed by the dropping of atomic bombs on Hiroshima and Nagasaki, a much cheaper alternative than fleets of aircraft using conventional explosives. On August 6, 1945, Assurbanipal's legacy of violence reached its apogee: 70,000 people in Hiroshima were killed in five minutes by one B29 aircraft carrying a single bomb. Colonel Tibbetts, whose crew dropped the weapon, looked down at his work and commented: "I couldn't see any city down there, but what I saw was a tremendous area covered by – the only way I could describe it is – a boiling black mass" (Dyer 1985, 96). Of the 76,000 buildings in the city, almost two-thirds were destroyed and the remainder seriously damaged. Three days later, a second atomic bomb was dropped on Nagasaki.

Since Hiroshima and Nagasaki, nuclear weapons have become many times more potent, and the world lived under the imminent threat of nuclear war until the early 1990s. Nuclear war, though less likely since the end of the Cold War, would be the ultimate domicide – the annihilation of the planet as the home of humankind (Fisher 1990, 123). If sufficient bombs were detonated, the resulting pall of black smoke would dwarf the one that resulted from the burning of Kuwait's oil fields in 1991. It would encircle the planet, block all sunlight, bring photosynthesis to an end, and, with the coming of "nuclear winter," consequently bring to an end all life on earth. This, of course, is the bang; the whimper of ecocide will take much longer.

Strategic Resettlement

Rural societies may escape bombing, but they experience domicide in other ways. A frequent phenomenon of war or its supporting activities is strategic resettlement from which, invariably, arises the loss of home. The motives for such resettlement include the need to remove civilians from battle zones or from the influence of opposing political ideologies. Uprooted villagers, forced into compounds, are demoralized by the change and thus easier to control. Meanwhile, their former homes are destroyed. During the latter half of the twentieth century, millions of Third World peasants were forcibly resettled in wartime from Afghanistan and Algeria to Vietnam and Zimbabwe.

The Algerian War of Independence from France lasted from 1954 to 1961, during the course of which several hundred thousand Algerians were killed. But "of all the hardships to which Algerian rural society was subjected between 1954 and 1961, none was more brutal than, nor had such far-reaching consequences as, the massive regrouping of the rural population by the French army partly to protect it but also to prevent it from actively assisting the guerillas" (Sutton and Lawless 1978, 332). Initially, the French army mounted a scorched earth policy, creating *zones interdites* (forbidden zones), which extended across the whole country. The inhabitants were expelled from these areas, which then became "free-fire zones." Abandoned settlements were destroyed by French troops, and no provision was made for rehousing the evacuees. This period was soon followed by a more systematic resettlement policy known as the Challe Plan (Talbott 1980, 186). By 1962, it was estimated that 3.525 million Algerians had been torn from their homes, which amounted to half of Algeria's rural population (Cornaton 1967, 61). Another French commentator called the operation "among the most brutal known to history" (Bourdieu and Sayad 1964, 12). It is significant that French critics used the word *deracinement*, which means uprooting. There was nothing civil in France's *mission civilatrice*.

As in Algeria, scientific advisers to the United States Department of Defense believed that "the solution in Vietnam" was to remove the rural people who formed the political base for the Viet Cong guerrilla forces (Ekberg et al. 1972, 110). Begun in the late 1960s, this "villagization" policy resulted in the destruction of 3,000 hamlets and the removal of their inhabitants to cities, camps, or "strategic villages" under government control. As a result, Vietnamese relocatees suffered effects ranging from psychological depression, apathy, and guilt to increased death rates from cholera, typhoid, and plague. They became "'ghosts – both living and dead' – wandering a land where everything that once made sense of their lives had been destroyed forever" (Isaacs 1983, 20).

The same "solution" was tried in separatist Eritrea by Haile Selassie of Ethiopia in the 1960s. Supervised by Israeli officers, Force 101 began in 1962 to repress the Muslims of the Eritrean lowlands. "At the end of the year, about three hundred villages had been razed, six hundred civilians massacred, and sixty thousand head of livestock slaughtered. Twenty-five thousand lowlanders fled to Sudan, and a policy of forced relocation was imposed on those who remained" (Harding 1993, 275). Things were no better under the dictator Mengistu in the 1980s, for hundreds of thousands of people, for war purposes, were forcibly resettled in the Ethiopian highlands.

A similar but better-known destruction and resettlement at the regional level occurred in Iraq in the 1980s. It is believed that about 3,000 Kurdish villages were destroyed and up to 500,000 persons relocated in order to crush the Kurdish rebellion and create a security zone along the Iraqi border with Iran. The Iraqi government denied that these were forced removals, stating that the relocation was "only a communal program aimed at uplifting economic and social standards of villages" (*Chicago Tribune* 1989). But under the umbrella of the "exclusion zone" created by the allies after the 1991 Gulf War, Kurds began to return to their home areas, rebuilding their towns stone by stone. In the words of one man: "I feel like I have just been born again ... Saddam, he even destroyed our dreams. Coming back is the only way to defeat him" (Nordland 1991, 29). When asked, in Michael Ignatieff's (1993) film *Blood and Belonging*, why he had returned to the ruins of the town of Halabja, an interviewee replied: "Our fathers and grandfathers were here; in all the world, there is nowhere sweeter; we have always lived here; it is beautiful, it is our place."

War-related relocations, often extremely brutal, were still ongoing in a number of countries as the twentieth century ended. Myanmar (Burma) is one of the worst offenders, as is Angola, most modern accounts of which use the word "displacement" with unpleasant frequency (Harding 1993). In the Western Sahara, Morocco has built a military Sand Wall over 1,000 km long across the country and driven well over half the inhabitants eastwards beyond it. Here, in Algeria, they are safe from napalm and phosphorus bombs, even if they have lost both their homes and their homeland. Nevertheless, the motto of the Saharan resistance is *Koul el watan aou shahada* (the whole homeland or martyrdom) (Harding 1993, 109). Within the Sand Wall, the Moroccan army and Moroccan settlers mine Western Sahara's resources, while indigenous towns like Tifariti stand bleak and empty.

There are no reports that peoples' lives have been significantly improved by war-related resettlement policies. Indeed, many have suffered both physically and psychologically, their situation made worse by the often feeble efforts to provide them with alternative living and working arrangements. At best, one may say that while their dwellings, neighbourhoods, and landscapes were destroyed, the evacuees retained at least their living bodies, though often with damaged minds.

The Creation of Military Installations

"If you wish for peace, prepare for war." Twentieth-century industrial nations took this Roman adage to heart in the creation of global networks of military bases, surveillance stations, and nuclear, chemical, and biological weapons-testing sites. The military's appetite for land is enormous; between 750,000 and 1.5 million square kilometres of public land are controlled by the world's armed forces. Often located in what appear to city-based military planners as "empty" areas, these installations actually occupy the ancient homelands of nomadic peoples such as Shoshones, Aleuts, Kazakhs, Uygurs, and Australian Aborigines.

Although the military installations of most industrial nations have been the generators of domicide, those of the United States provide some of the most spectacular and accessible cases. Nuclear weapons testing, for example, has not only devastated the homelands of desert nomads, but it has also rendered uninhabitable a number of remote Pacific islands. During the fifty years 1946–96, scores of nuclear bomb tests were conducted on Pacific islands by the United States (1946–62 on Bikini, Enewetak, Johnston, and Christmas), Britain (1957–68 on Christmas and Malden), and France (1966–96 on Mururoa and Fangataufa).

Americans used Bikini atoll in the Marshall Islands for the testing of twenty-three atomic bombs between 1946 and 1958. Islanders were told that the project "would benefit mankind" (Ellis and Blair 1986, 814). Yet, scientists at Bikini deliberately detonated their Bravo "device" upwind of Rongelap and Utirik atolls, the worst exposure of human beings to radiation since Hiroshima and Nagasaki. The 166 Bikini islanders were evacuated in 1946, returned, removed from their atoll again in 1956, then sent back in 1971 when United States authorities deemed the islands to be safe. Seven years later, the same government declared the atoll "uninhabitable," and again removed 167 Bikinians. The 142 people of Enewetak were similarly treated. Its level of radioactivity was so high that Runit Island, Enewetak, was completely cemented over with a concrete cap. It was meant to contain the lethal radioactivity for 50,000 years, but is already cracking. So contaminated were the atolls of Rongelap and Utirik that their whole populations were removed. Returning later, the Rongelap people found their island too contaminated, and they were again evacuated in 1985 by Greenpeace activists.

Deprived of their homelands, islanders who were relocated elsewhere became demoralized and apathetic, suffering not only from radiation-induced illness but also a variety of psychiatric disorders and social problems such as family disintegration and loss of cultural skills. Having totally lost their land-based economic independence, the islanders of the four "radiation atolls" have been reduced to a life of food-aid welfare. They have received nearly $100 million in compensation, but that has not eased the spiritual pain caused by loss of homeland. Moreover, their radical change in diet has led to numerous medical problems. But mere numbers, which are small, cannot tell the whole story; Marshall Islanders call the American lists of statistics "people with the tears wiped away."

Also in the Marshall Islands, the atoll of Kwajalein was chosen to be the bull's eye for testing long-range ballistic missiles launched from

Vandenberg Air Force base in California. Establishing a typical American base on the atoll, the military authorities felt the need to forcibly remove over 8,000 Kwajalein islanders to the nearby island of Ebeye. Formerly inhabited by a few fishermen living in thatched huts, the sixty-six acres of Ebeye were soon covered with rows of cinder-block apartments into which 6,000 Micronesians were crowded, with up to thirty to forty people per apartment, sleeping and eating in shifts. A further 2,000 islanders were compelled to live in shacks on the beach. By the 1990s, over 9,000 people were crammed onto the island, which was supported by water and sewage systems designed for 3,000. The new inhabitants of Ebeye have been called "compensation junkies, greedy, irresponsible, weak-livered, money-grabbing parasites" (Nakano 1983, 175). Ebeye's vegetation has disappeared, the pristine lagoon is now a cesspool, diseases such as tuberculosis, malaria, and dysentery are rampant, and islanders who recently were self-sufficient must now import 98 per cent of their food. The health and homes of the islanders have clearly been sacrificed to US military aims. Meanwhile, across the strait, American forces personnel clock missiles that roar in, then go off to play golf in a country-club setting.

Such island histories are encapsulated in the haunting novel *Greenvoe* in which George Mackay Brown describes what happens when an Orkney island is taken over by a secret military installation known as Black Star. Public interest rhetoric is evident; Black Star is utterly essential to the security of the Western world. It does not matter whether the inhabitants comply or try to resist: "The houses of the village went down, one after the other ... they collapsed before clashing jaws and blank battering foreheads" (Brown 1972, 217). Once removed from the island, the former inhabitants' only means of resistance is to creep onshore in the dark of night to enact place-based harvest rituals going back to prehistory.

But going back is impossible when people are removed hundreds of kilometres from their island. The fate of the islanders of the Chagos Archipelago (Diego Garcia) in the Indian Ocean is not well known if only because many attempts have been made to suppress this knowledge. The need for a strategic outpost in the Indian Ocean is a major component of Western military strategy. In 1965, the colony of Mauritius declared its right to become independent. The British agreed but only on condition that Mauritius gave up its claims to the Chagos Archipelago, 1,800 km north, in perpetuity (Alladin 1993). This island group then became a new colony – the British Indian Ocean Territory (BIOT). Britain then signed an agreement leasing BIOT to the United States for seventy years without payment. In return, the British were allowed to purchase Polaris nuclear missiles at discount rates. The

British government had assured Mauritius that Diego Garcia, the southernmost island of the Chagos group, would not be used as a military base.

Whereupon the United States Navy took over Diego Garcia and emblazoned a water tower near the harbour entrance with the slogan "Welcome to the Footprint of Freedom," an allusion to the island's shape. Only there was little freedom, because BIOT then became the only British colony that cannot be visited legally by ordinary British civilians (Winchester 1985, 39). When BIOT was formally established, the Chagos Archipelago had 2,000 islanders who worked for a French-run coconut oil factory. They lived a pleasant, relaxed, and moderately prosperous life in a series of attractive villages. They were British citizens entitled to the protection of Her Majesty's government.

None was given (Madeley 1985). The US Navy insisted that the whole area should be made "sterile," that even islands 150 kms north of Diego Garcia should be "swept clean." The British replied that the islands contained only some "rotating contract personnel" who would soon be resettled. To speed up the process, the British government killed the local economy by buying up the coconut oil factory and closing it down. Islanders' jobs vanished at a stroke, the money economy collapsed, and boats ceased to call. Slowly, the "small migratory population" was cleared and dumped in Mauritius, in a slum near the Port Louis docks.

Paul Moulinie, entrusted by the British government with the unpleasant task of winding down the copra industry and evicting the islanders, said: "We told them we had orders from BIOT ... we are closing up. They didn't object. But they were very unhappy about it. And I can understand this: I'm talking about five generations of islanders who were born on Chagos and lived there. It was their home." An islander responded: "We were assembled in front of the manager's house and informed that we could no longer stay on the island because the Americans were coming for good. We didn't want to go. We were born there. So were our fathers and forefathers who were buried in that land" (Winchester 1985, 41). The islanders, of course, were not consulted. No compensation was offered until the story became known worldwide. Resigned, without resistance, these contaminants of a pristine defence environment took their meagre compensation and tried to fit into the alien life of Mauritius. According to the Mauritian government, in 1981, 77 per cent of the Ilois wanted to return to Diego Garcia (Instituto del Tercer Mundo 1997, 388).

Even in 2000, more than a generation after their removal, many Ilois still wished to return to Diego Garcia. Although they are now citizens of Mauritius, they retain their distinctive "Chagosien" culture and in

1998, succeeded in being recognized by the United Nations as an "indigenous community" with a special relationship to the islands of the Chagos archipelago. Unfortunately, the American military shows no signs of abandoning Diego Garcia. In the *Atlas of Mauritius* (Saddul 1996, 28), the island's western half shows three settlements: the "US Military Base," being accompanied, ironically, by Splendidville and Eclipse Point. The three settlements of the eastern half of the island are designated "ruins."

Fortunately, after thirty years of pressure, the islanders may be about to return. In November 2000 a London High Court ruling declared that the Ilois were unlawfully evicted by British diplomats, who had referred to them as "a few Tarzans or Men Fridays" (MacAskill 2000, 1). Anticipating the recommendations of chapter 7 of this book, the Ilois leader declared: "We ... insist that we are consulted during the process leading to our return. We must never again rely on governments to tell us what we should or should not have" (Bancoult 2000, 18).

For cynicism, cruelty, mendacity, and inhumanity, the saga of war-related domicide is difficult to surpass. It is a bleak picture of callous insensitivity on the part of the powerful toward those considered expendable or "in the way" of their grand projects. Whether battles are won, territory controlled, or strategic plans realized, the benefits accrue only to the perpetrators.

Fortunately, wars come to an end, and bases are relinquished. One of the most characteristic features of war-related domicide is that the victims generally try to return home as soon as possible in order to reconstruct their dwellings and their lives. Even Hiroshima and Dresden have been reconstructed, and in terms of the restitution of identity, it is remarkable how much effort has been made to reconstruct traditional town centres in loving detail, the best-known example being the old town of Warsaw destroyed by the Nazis in 1944. In this way, the "lasting wounds on the human meaning that has been accumulated over centuries" (Violich 1993, 11) may in part be treated. Thus, we learn that: "Except in brief moments of crisis, human survival is never just individual biological persistence, but the need to have a communal place or to re-establish the continuity of past places ... Somehow, in extremity, one discovers that the intersubjective reality of place has a more general and fundamental human significance than objective form and function or measures of the material setting" (Hewitt 1983b, 277). Ironically, those who suffered most generally have the least say in the reconstruction efforts. Enemy bombing provided a free slum clearance service, creating space for politicians and planners to realize their dreams.

COLONIAL GEOPIRACY

"Imperialism" means the practice, the theory, and the attitudes of a
dominating metropolitan centre ruling a distant territory; "colonialism,"
which is almost always a consequence of imperialism, is the implanting
of settlements on distant territory.

Edward Said (1993, 9)

The year 1492 saw the beginning of a European imperialist project that
forever changed the face of the globe. Imperialism is "an act of geographical violence" through which virtually all the world's spaces have
been explored, mapped, and brought under control (Said 1993, 225).
The colonizer's model of the world (Blaut 1993) sees the whole globe
as open to conquest. By 1914, about 85 per cent of the earth's surface
was under the domination of European powers.

The history of the Old World was a story of endless clashes
between nomads and settled people, between the steppe and the
sown. When agrarian Europeans invaded the New Worlds of the
Americas and Australasia they found what they regarded as *terra nullius* (empty lands) sparsely inhabited mostly by nomadic peoples.
They regarded nomads as inferior, as being at a lower stage in the
development of civilization. To take the land of the nomads, to settle
them in reserves, to "civilize" them, and to turn steppe into sown
land was to the European invaders not merely a satisfaction of their
land greed, but also part of the great "civilizing mission" that they
had taken upon themselves as the "white man's burden." They ratio-
nalized that the ways of the natives were being changed for their own
good. Thus, from 1492 until the twentieth century, common good
rhetoric aided and abetted the theft of three continents. Essentially,
similar processes of dispossession took place in the Old World – on
the Russified and Sinicised fringes of the Russian and Chinese
empires in Siberia, Mongolia, and Central Asia. Africa, too, endured
a phase of white settlement, most notably in Algeria, Kenya, South
Africa, and Rhodesia/Zimbabwe.

Such colonization can be called geopiracy, the forcible theft of the
territory of others. It does not necessarily involve warfare, for blandishments and guile might be sufficient to wrest land "title" from
indigenous people who typically knew no such concept. Typically, aboriginal populations were first decimated by disease and thoroughly
demoralized by missionaries. Those who remained were forcibly
rounded up and confined to reserves, often on inferior land with few
resources. In this way, their homelands were actually rendered empty,
to be immediately filled by waves of immigrant settlers.

Two rather obscure cases from the mid-nineteenth century can be regarded as typical. The indigenous Ainu of northern Japan were severely encroached by the Japanese settlement frontier moved northward in Hokkaido. By means of laws enacted by the Japanese, the Ainu were deprived of much of their land, which at that time was a bountiful forest. Japanese agriculturalists took over land suitable for crops, while the forests were cut down "for the profit of 'the nation of Japan' and the corporate giants" (Kayano 1994, 9). Deprived of their land base, the Ainu lost their traditional crafts and rituals as well as their hunting and fishing way of life. They were then subjected to "Ainu hunting," whereby Japanese labour recruiters forcibly took the people of Nibutani and other villages to work in factories in distant towns. The Ainu story is one of "prejudice and discrimination, dislocation and lost lands, forced labour, increased illness and death, and environmental destruction" (Cybriwsky 1995, 230). Of the 300,000 Ainu of 1800, less than 25,000 now remain in Hokkaido; other estimates are even lower (Rudnicki and Dyck 1986).

Geopiracy on Easter Island had a much worse result (Porteous 1981). By the early 1860s, a combination of slave raids, disease, and evacuation encouraged by missionaries and entrepreneurs had reduced the Rapanui population from several thousand to a mere 110. Armed with "land titles" on which some natives had been induced to make their marks, the buccaneering Jean-Baptiste Onexime Doutrou-Bornier soon took control of the whole island. Doutrou-Bornier and the missionaries succeeded in rounding up all the remaining islanders and confining them to the single village of Hangaroa. The rest of the island became a sheep ranch. The island was then formally declared a *res nullius* (uninhabited zone), by Chile, which annexed Easter Island in 1888. At the end of the twentieth century, the settlement pattern had changed very little, though the Rapanui population now exceeds 3,000.

Decimation, roundup, confinement, and use as labour when required – this is the fate of indigenous peoples the world over. The result is loss of homeland and the disintegration of identity. Some of the most throughly documented cases of geopiracy occurred in the Americas. Although usually regarded as empty lands by Europeans, the Americas actually contained one hundred million native Americans in 1492 – about a fifth of the human race. Ronald Wright's *Stolen Continents* (1992) sums up the colonization process in the title's two emotive words.

United States

The story of the aboriginal peoples of what became the United States is one of an appalling loss of homelands. Aboriginal people were driven off their land and treated as foreign nations, subjects, or wards (Jackson 1993, 2). The process began in earnest in the eighteenth century with a blaze of self-justificatory rhetoric. In 1792, George Washington, the president of the United States, addressed the hostile Delaware nation, among others, in this manner: "Brethren: the President of the United States entertains the opinion that the war which exists is an error and mistake on your parts. That you believe the United States wants to deprive you of your lands and drive you out of the country. *Be assured that this is not so;* on the contrary, that we should be greatly gratified of imparting to you all the blessings of civilized life; of teaching you to cultivate the earth, and raise corn; to raise oxen, sheep, and other domestic animals; to build comfortable houses; and to educate your children so as to ever dwell upon the land ... *Remember that no additional lands will be required of you, or any other tribe, to those that have been ceded by former treaties*" (Jackson 1993, 40). The Indians were urged to abandon their cultures, assimilate European ways, and to make a peace "founded on the principles of justice and humanity."

Justice and humanity, apparently, did not include the right of aboriginals to remain on lands they had used ancestrally. City-bred notions of "the noble savage" were anathema to settlers and business people who saw the Indians' low-density use of vast areas of land as an impediment to their conceptions of progress and profit. From the mid-eighteenth century, a great wave of European settlement rolled westward from the eastern seaboard, uprooting Indian nations and driving them ever west and south into steppe lands and deserts.

The Delaware, for example, who received the glad tidings of justice and humanity noted above, were compelled in 1817–18 to cede title to their lands. By the 1830s much of the land northeast of the Mississippi had been cleared of Indians. Significantly, it was the secretary of war who congratulated his countrymen in 1833 on the fact that "the country north of the Ohio, east of the Mississippi, including the States of Ohio, Indiana, Illinois, and the Territory of Michigan [had been] cleared of the embarrassment of Indian relations." To which the commissioner of Indian Affairs added that it was "grateful to notice" how well the Indians' condition had been ameliorated under the policy of removal; for they were now "dwelling on lands *distinctly* and permanently established as their own" (Jackson 1993, 49).

Further south, the Cherokee suffered a fate that, like Hiroshima, is too well-known to rehearse here (Wright 1992, 207). "Build a fire under them," advised President Andrew Jackson. "When it gets hot enough, they'll move" (Carter 1976, 83). In 1838, the United States Army rounded up the Georgia Cherokee and kept them for several months in disease-ridden camps. Then they forced them to trek, at bay-onet point, throughout the whole winter across 1500 km of rough ter-ritory to Oklahoma. One-quarter of the 15,000 Cherokees died along this Trail of Tears. John Burnett, a soldier involved in this forced removal, later reflected: "School children of today do not know that we are living on lands that were taken from a helpless race at bayonet point to satisfy the white man's greed" (King and Evans 1978, 180; King 1979). Such is memoricide. The Cherokee nation addressed the United States Congress as follows: "You abolished our government, annihilated our laws, took away our lands, turned us out of our hous-es, denied us the rights of men, made us outcasts and outlaws in our own land, plunging us at the same time into an abyss of moral degra-dation, which was hurling our people to swift destruction" (Reed 1979, 158)

By 1893, much of their Oklahoma land was opened up to white set-tlers. Even before this outrageous final solution, Helen Hunt Jackson had published her best-selling *A Century of Dishonor* in 1881. But per-haps the words of Sioux chief Sitting Bull best sum up the motives of the whites: "The love of possession is a disease with them" (Turner 1974, 255).

As the settlement frontier moved west, nation after indigenous nation was forced to move westwards before it or be packed into east-ern reservations that were infinitesimally small. And removal policies were just as readily applied on the western side of the Mississippi, as the fate of the Pawnee attests. The Pawnee nation had evolved an envi-ronmentally sound way of life in the Great Plains based on horticulture and the hunting of bison. After 1830, great pressure was placed upon them to renounce their culture in favour of Euro-American concepts of civilization. Like the Cherokee, they were expected to become yeoman farmers and suddenly develop a sense of private property. Settlers pushed the Pawnee from their lands, and the Indians fought back by raiding farms and making off with livestock. By 1856, settler attitudes had become very blunt: "Pawnee Indians are in possession of some of the most valuable government land in the Territory [of Nebraska]. The region of the country about the junction of the Salt Creek and the Plat-te is very attractive, and there would immediately grow up a thriving settlement were it not for the Pawnees. It is the duty of Uncle Sam to remove the Pawnee population" (Wishart 1979, 392). Uncle Sam responded with alacrity.

The Pawnee were stripped of their land and provided with a small reservation, annuities, farm equipment, and promises of protection. This was merely a palliative, for they were again engulfed by the tide of settlement. Within the reservation, conditions soon proved intolerable, and the structure of Pawnee society broke down. Settlers invaded the reservation for timber, and the district court in Omaha ruled in 1872 that no redress was possible because the United States had no jurisdiction over crimes committed on Indian reservations.

Even at this late date, the Pawnee wished to stay in Nebraska, reluctant "to sever the bond with the homeland that contained their sacred places and the graves of their ancestors, and which was the setting for their traditions and world view" (Wishart 1979, 399). Their attachment to the land was broken only when the Pawnee realized that their traditional way of life could not be maintained on such a small reservation. Taking an enormous gamble, they decided to move south to new lands, ahead of the settlers, in order to preserve their culture. Yet, on their new reservation in Indian Territory, their population declined rapidly, and they soon regretted their decision to migrate. Yearning for their lost homeland, the Pawnee all but disappeared as a distinct group.

Other nations fared little better. In 1863, over 8,000 Navajo were rounded up and led on an exhausting 750 km trek to what is now New Mexico. Ironically, in 1974, over 6,000 Navajo were evicted from this new reservation because of a dispute with the Hopi (Scudder 1982, ix). Unlike the Pawnee, however, the Navajo nation has survived rather well.

In 1881, Jackson provided a summary statement of the fate of Indians in the United States, which remains valid after yet a second century of dishonour: "It has come to be such an accepted thing in the history and fate of the Indian that he is to be always pushed on, always in advance of what is called the march of civilization, that to the average mind, statement of these repeated removals comes with no startling force and suggests no vivid picture of details, only a sort of reassertion of an abstract general principle. But pausing to consider for a moment what such statements actually mean and involve; imagining such processes applied to some particular town or village that we happen to be intimately acquainted with, we can soon come to a new realization of ... such uprooting, such perplexity, such loss, such confusion, and uncertainty" (Jackson 1993, 64). Some see this as attempted genocide; it was certainly domicide achieved on a grand scale.

Canada

In contrast to her execration of US policies, Helen Jackson extolled Canada's relationship with its indigenous peoples, which rarely involved American-style Indian wars, so "needless and wicked." Euro-Canadians, she believed, "seldom remove Indians" (1993, ix). British policy was certainly more pro-native than in the United States. Indeed, there was much less contempt, less overt cruelty, and few long-distance trails of tears in Canada. Yet it is probably true to say that all of Canada's Indian groups have been compelled or encouraged to either relocate or to abandon their wide ranges for squalid concentrations around a fort, trading post, or missionary station.

Canada's late nineteenth-century settlement of the huge tract of prairie and forest land between the Great Lakes and the Rockies was quite different from the prairie conquests of the United States. In the 1870s, the United States was spending $20 million a year on Indian wars alone, while Ottawa's entire budget was only $19 million. Smugly, the Canadian government negotiated treaties in advance of white settlement. Native peoples were induced to sign these treaties by missionary pressures, as a result of fear of what was happening south of the border, and because of the rapid depletion of their staple, the bison herds. "Behind the eagerness of some Indian leaders to negotiate lay a recognition that their way of life was collapsing" (Miller 1989, 164).

Across the country, the Canadian native population dwindled because of loss of culture, health problems, poor diet, inadequate housing, low incomes, and white hostility. Only in the 1930s was there evidence that the population of status Indians had begun to climb again. Gathering strength after the Second World War, Canada's First Nations began to organize against unfair treaties, racial discrimination, and unjust treatment. A rallying call came with the federal government White Paper of 1969, which, ignoring the long history of white inhumanity, claimed that the Indians' chief problem was their legal distinctiveness. The policy recommended the abolition of Indians' special status and their assimilation into the Canadian mosaic.

Since that time, Canadian native peoples have chosen to assert their distinctiveness and have fought against the exploitation of their remaining lands for industrial resources. They were especially incensed to discover that South African officials had studied Canadian Indian reserves as a model for the policy of apartheid. Existing aboriginal rights were not entrenched by constitutional amendments until 1982. Elsewhere, in large areas in which the land was never surrendered by treaty – much of Labrador, Quebec, British Columbia, the Yukon, and the Northwest Territories – Indian nations began to dispute the white assumption that Indians had no title whatsoever to lands they had used from time immemorial.

In British Columbia, to take a single example, the Nisga'a were expelled from their homeland in the Nass Valley from 1883 onward to make way for white miners, loggers, and commercial fishermen.

Despoiled of their original territory of about 24,860 square km, and having signed no treaty, they took their case to Ottawa and then to London in 1906 and 1909. In 1915, they were rewarded with a token land grant of 76 square km, about 0.3 per cent of their former territory (Raunet 1984, York 1990). The contemporary provincial government's case was put forward in a series of platitudes by its agent, a Mr. Planta, who told the Nisga'a that all their land was now crown land and that it would hand over to the federal government "such tracts of land as may be deemed reasonably sufficient for all the purposes of the Indians. The sole project of the Government being to protect the Indians in their purposes and property" (Raunet 1984, 89). On hearing such casuistry, an old, blind Indian leapt to his feet and cried: "Who is the chief that gave this land to the Queen? Give us his name, we have never heard it."

The descendants of the dispossessed Nisga'a continued to fight their case by legal means, and they were encouraged in 1991 when the British Columbia provincial government at last recognized that aboriginal rights in BC had not, in fact, been extinguished. Negotiations ensued, but they were delayed by corporate interventions. In an open letter to the people of British Columbia, Joseph Gosnell Senior, president of the Nisga'a Tribal Council, declared:

We have been sitting at the negotiating table for nearly two decades. A generation of Nisga'a men and women has grown old at that table. And suddenly, at a critical stage in our negotiations, when an agreement-in-principle is within reach, the backlash has begun. In this campaign, we see the hand prints of powerful vested interests. We believe they are trying to derail the talks which threaten to interrupt their unfettered plunder of our territory's precious resources. They have one goal: to intimidate politicians into scuttling Nisga'a and other aboriginal negotiations. It would be a mistake to confuse these vested interests with the common good. While they have systematically stripped Nisga'a lands of our fish and forests – at handsome profits – the Nisga'a have received little or no benefit. First Nations are growing tired of trying to educate and explain – again and again – the history of our brutal treatment at the hands of explorers, colonizers, and, now, the faceless number crunchers at big corporations (Gosnell 1994, A5).

On February 15 1996, the governments of Canada and British Columbia signed an agreement in principle with the Nisga'a, and the ratification process for the treaty was completed and given Royal Assent on April 13, 2000. The Nisga'as are to receive about $200 million, ownership over 1,930 square km of land, and the right to self-government at a quasi-municipal level. The wheels of justice grind very slowly indeed. In North American terms, the Nisga'a may be regarded as quite successful; after 110 years, they at last have gained jurisdiction over almost 8 per cent of their ancestral homeland.

But others were not so lucky, and dispossessions have not ceased. The Lubicon Cree in Alberta, for example, never signed a teaty and so were never assigned a reserve. Their traditional lands were in the middle of Alberta's oil patch and were invaded by roads to facilitate oil and gas extraction. Large-scale logging is completing the ruin of the Lubicon homeland. In 1989, the United Nations Human Rights Committee found that it was impossible for the Lubicon to gain redress through the Canadian legal system. In fact, the 500 remaining Lubicon have been told that it would be better for them to move and amalgamate with other Cree bands elsewhere. Such administrative removal and band amalgamation is quite common in Canada. In 1955, the Ojibway of White Dog, Ontario, were joined by two other bands. That from One Man Lake had been moved because its homes and burial grounds had been flooded by a hydroelectric project. The Swan Lake band was moved because the government refused to build a school for only a handful of families. Forty years later, the White Dog still resent the disruption forced upon them in negotiations between chiefs who spoke no English and an Indian agent who spoke no Ojibway. The chief of the former One Man Lake band expresses the problem thus: "Because we were forced here and exiled from our own lands, there was never really acceptance to the community. There was a feeling of intrusion. We had no choice in the matter. We had no choice but to move here" (Gray 1997, A6). Refusal of the government to build a school also forced the people of Yuquot on Vancouver Island to move to a site provided at Gold River in 1967. This site was adjacent to a pulp mill, which caused so many health problems that the band was moved again, in 1996, to a healthier spot. Such administrative removals still continue, causing hardship and misery.

Perhaps the most controversial case of the 1990s was that of the Innu, who inhabit a vast area of Labrador and adjacent parts of Quebec. The Innu are hunters and require large tracts of land to support their way of life. Until about 1950, their culture remained intact: they spent winters in coastal villages but moved out annually from these into *nutshimit* (the country), where they lived in tents and ate the animals they hunted.

From the 1950s, however, both missionary and government pressures compelled the Innu to gather into permanent coastal villages. Once children were in school – organized insensitively according to a southern timetable – the traditional culture collapsed, for the long school year rendered it no longer possible to make the annual trek into the wildlife-rich barrens. Children became alienated from their parents, and the Innu in general were encouraged to think that their culture was worthless. Like all indigenous cultures, Innu culture could not be taught in school; it could only be lived (Wadden 1996).

The Innu came to the world's attention in several ways. First, in 1965, several hundred Innu were moved twice before they were persuaded by missionaries and the Newfoundland government to form a permanent village on an offshore island known as Davis Inlet. Significantly, they named the village Utshimassit (the Place of the Boss). Although they were promised heated houses, running water, and a sewage system, none of these emerged. Because permanent settlement reduced both the spatial range of, and the time available for hunting, the Innu deteriorated. Alcoholism, family breakdown, and high teenage suicide rates became endemic, the latter becoming a media issue in 1993. Shamed in the international press, the Newfoundland and federal governments included Innu leaders in the planning of a $180 million new village at Sango Bay on the mainland. With better fishing and ready access to the caribou herds, Sango Bay should have been chosen in the first place.

The sedentarization of the Innu, however, enabled the Newfoundland government to claim that the interior of Labrador, mostly crown land, was essentially empty. By 1995, a frenzy of speculation had resulted in 280,000 mineral rights claims staked throughout over half of the Innu nation's territory (Hildebrand 1997). As Daniel Ashini explains: "Mining companies and other industrial interests occupy our land as if we don't exist" (1996, vii).

To add injury to injury, the third assault on the Innu came as a takeover of the airspace of their homeland. Centred on the Canadian Air Force base at Goose Bay, over 100,000 square km of Labrador airspace has been reserved for military use (Thomas 1995, 63). In this area, NATO jets fly up to 40,000 sorties per year. Many of these are low-level training flights flown at supersonic speeds. The deafening sonic booms cause waterfowl to desert their nests, mink and foxes to devour their young, and caribou to stampede. In other words, the unbearable noise, which the United States Air Force typically calls "the sound of freedom," destroys the material basis of Innu culture.

Driven from their homeland by the military, missionaries, mining companies, and politicians, the Innu clearly see that without a large land base on which to hunt, their culture is doomed: "If the land is gone, there is no culture" (Wadden 1996, 162). As Innu Roxie Gregoire states: "The government that invaded our country now thinks we are weak enough to bury alive. The one thing that has stopped our complete breakdown as a people has been the months we still live away from the village in our tents in the country." Her sister Tshaukuesh, thrown in jail for protesting the takeover of her country, asks of Canadian politicians: "What kind of a people are they? They have no heart" (Wadden 1996, 162).

In his examination of the relocations of Indians in Manitoba and Saskatchewan, Dickman (1973, 169) suggests two rather obvious but usually ignored conditions for any relocation. First, no relocation should occur unless those to be relocated accept the change as a desirable move. Second, adequate resources must be made available so that a successful change can occur. As a former Indian Affairs minister is quoted as saying: "Many of our imposed solutions of the past have not been successful, to put it mildly" (Bronskill 1993). A spate of recent books provides details of many more cases of white duplicity in the theft of Indian lands in Canada. Their very titles form a significant processual sequence: *Without Surrender, Without Consent* (Raunet 1984); *The Dispossessed* (York 1990); *Skyscrapers Hide the Heavens* (Miller 1989); and *Maps and Dreams* (Brody 1988).

International Implications

Whether in Canada, Australia, or the United States in the nineteenth century, or in the tropical forests of South America, Africa, and Southeast Asia today, colonization and economic exploitation result in geopiratic domicide – large-scale removals, disempowered victims, loss of identity, and often a common good rhetoric as justification. The motives are economic and political, and the winners are Western and Asian commercial interests. Raunet has summarized the dilemma of the Nisga'a, and all such dilemmas, as follows: "The Native fight is the fight of all those trying to regain a sense of owning their own lives, those who are threatened by the 'machine': land-deprived peasants of the Third World, citizens of the industrial heartlands refusing atomic plants or nuclear weapons on their doorsteps, fishermen against pollution and the asphyxiation of the sea, workers victimized by inhuman production and investment plans – opponents of all the madness of the present age" (1984, 236). Hugh and Karmel McCallum (1975, 2) have also expressed the meaning of loss of the land that is home to native peoples: "The closest definition we can come to is that, for Natives, land is for use; it is like a Mother. It is a breadbasket, protector, and friend. It is something you live with easily, you don't fight. It is something you cherish and return to when you are sick, frightened, or lonely. It has always been there, and it always will be there. And out of it comes your being, the reason for your existence, the only power you have in a white man's world. If you lose it or sell it or have it taken away from you, then you are dead or, at best, a second class white man" (1975, 2).

Since the 1970s, these ideas have at last gained some credence in white societies, and some weak attempts are being made to reverse colonial geopiracy in Hawaii, New Caledonia (Kanaky), and notably in New Zealand. In 1863, the *New Zealand Settlements Act* confiscated three million acres of Maori lands; an agreement signed in 1995 returned 2 per cent of this land (60,000 acres) to them. And only in 1992 did the *Mabo* judgement force the Australian national government to abandon its stand that Australia was *terra nullius* (empty lands) when the first Europeans landed, thus opening up the issue of Aboriginal rights to negotiation.

Those of us prospering in the "white colonies" today are benefiting from land theft on a massive scale. Were acts that occurred less than a century ago being enacted today, we would rightly charge the perpetrators with crimes against humanity. But such crimes are still being committed in the forests and steppe lands of the Third World by modernizing Third World elites that hold attitudes typical of white settlers in the nineteenth century: natives are a different order of beings; that they and their cultures are inferior to the majority culture, which is city-based; they stand in the way of progress, which means profits for the elite; and, it would be in their best interests to be stripped of their culture and land and be modernized. The peoples affected are the world's remaining hunters, gatherers, shifting cultivators, and herders.

The sedentarization of nomadic herders has been attempted worldwide, with equivocal results. In the 1920s, totalitarian states were particularly interested in taming nomads. In a cruel war in their Libyan colony, the Italians attempted to take grazing land for their own settlers by cementing up wells to discourage Bedouin flocks from using them and to deprive them of food. Forced collectivization of herds in Mongolia and Kazakhstan caused some herders to flee to other countries. The forced sedentarization of nomads in Iran by Reza Shah in the 1920s stored up trouble for his descendants in the revolutionary 1970s. Post-Second World War attempts to form grazing co-operatives in Somalia, Kenya, and elsewhere led to many failures because of nomad resistance to being organized by external authorities. Attempts to turn nomads into agriculturalists are usually misguided, although Ebrahim (1984) reports Saudi Arabia's considerable success in convincing nomads to become industrial operatives. He notes that whereas voluntary sedentarization usually works well, induced settlement is not appreciated, and forced settlement is usually catastrophic. The Tuareg, for example, have resisted such attempts, resorting to armed rebellion during the 1980s in Mauritania, Mali, Niger, and Chad.

Thus, while the Western world's native peoples are fixed in space, or even expanding their territorial control, the developing world's remaining nomads are threatened on all sides. Nomads do not fit easily into national boundaries, and they look and behave very differently from majority populations. Modernizing elites see nomads as unmodern, unproductive, wasteful of land, unsubmissive, difficult to enumerate, and as an embarrassment to the developing state. Vast areas of land have therefore been taken from nomads by regimes as various as Kenya (capitalist), Tanzania (socialist), and China (communist). The rationale is heartbreakingly familiar. The Sudanese Arab government, keen to sedentarize black nomads in the south of the country, declared: "Sedentarization ... is a means of improving the economic and social conditions of those communities ... to integrate them into the life of the nation, and to enable them to contribute fully to national progress" (Ellwood 1995, 8). The Israeli excuse for sedentarizing their Bedouin is depressingly similar: "We want, as a democratic government, to give all citizens the modern services that a state should give its citizens" (Ellwood 1995, 9).

The chief crimes committed by nomads in the modern state are that they are difficult to control and that they take up space that could be put to uses favoured by central governments wedded to large-scale agriculture, forestry, and mining. Given current pressures, nomads will be largely wiped out or sedentarized early in the twenty-first century, though small numbers may be retained in "environmental zoos" as tourist attractions. It is the goal of Survival International (Hanbury-Tenison 1991), Cultural Survival, the Minority Rights Group, the World Council of Indigenous Peoples, and similar organizations to prevent this process.

In 1998 the World Conservation Union released its Red Book of endangered plant species, product of a 20-year study. One in eight of existing plant species are in danger of extinction, mainly because of "loss of habitat" and "competition with nonnative species." The analogy with indigenous peoples could not be more clear.

Although Third World governments habitually ill-treat their indigenous peoples, in the Western world, the situation in the 1990s is that the legal concept of *terra nullius* (empty lands), the dubious basis of white dispossessions, has been widely rejected. Concepts of ethnic assimilation have given way to multiculturalism. Connection to the land is sought by white society, which now looks to native peoples for environmental wisdom. Both cause and consequence of this change in viewpoint is the massive resurgence of Aboriginal land claims, which currently cover almost the entire land surfaces of Canada, the United States, Australia, and parts of Africa, notably South Africa (Cant et al. 1993, Christopher 1994). These claims will be strongly resisted, and thus the potential for future conflict remains acute.

RESETTLEMENT PROJECTS

Landscape ... may be conceived as the outcome of a struggle among conflicting interest groups seeking domination over an immediate environment.

Ghazi Falah (1996, 256)

The struggle to establish mastery over territory is often waged in terms of ideology, whereby ideologies become manifestations of authority (Baker and Biger 1992). It is impossible to draw a hard-and-fast line between strategic resettlement in war, discussed earlier, and ideologically driven resettlement plans that occur in both war and peace. In general, however, we may distinguish two broad resettlement categories: ethnic cleansing – often based on racial, cultural, or religious grounds, and which may involve warfare; and settlement rationalization – which usually occurs in peacetime and of which the motives are political and administrative. While the motives for ethnic cleansing may be clear to all concerned, people relocated politically or administratively often do not understand the reasons for their resettlement. And in both cases, the victims of resettlement often disagree violently with both rationale and process.

Ethnic Cleansing

Although the term is of recent coin, the process is as old as civilization. The essence of ethnic cleansing is the targeting of specific population groups for elimination or expulsion from a particular site or region (Bell-Fialkoff 1993). In Rwanda, Burundi, and East Timor in the 1990s, for example, elimination and expulsion went hand in hand. The ultimate aim is to "purify" space of those regarded as aliens or impure. In some cases, as in South Africa and Guatemala, those to be eliminated, expelled, or subdued form the majority of the nation's population. Defenceless civilians are prime targets, and atrocities such as massacre and mass rape are not uncommon. Those left alive are deported, losing their homes, neighbourhoods and homelands. In some cases, the victims' dwellings are taken over by the perpetrators; in others, they are destroyed. Colonial geopiracy could well be regarded as a form of ethnic cleansing.

Ethnic cleansing was extremely common in the twentieth century, especially in the Middle East and Africa. The Armenians of eastern Turkey were either exterminated or exiled in 1915. Today, the Turks are extending this policy to the Kurds and Suriani Christians (Dalrymple 1997, 113), while Armenians, now with a country of their own, practise ethnic cleansing on the Azeris of Nagorno Karabach. In the early 1920s, Turkey and Greece forcibly exchanged large numbers of people. Famed Greek writer Nikos Kazantzakis has dramatized the plight of those Greeks forced to leave Turkey and, in doing so, draws attention to the importance of lost land and ancestors: "Uproot yourselves! The strong of the earth command. Go! The Greeks to Greece, the Turks to Turkey. A lament rose throughout the village; the people ran back and forth in confusion, bidding farewell to the walls, the looms, the village spring, the wells. They went down to the seashore ... and chanted dirges. It is difficult ... for the soul to tear itself away from familiar soil and familiar waters" (Kazantzakis 1974, 12–16). At the cemetery, the oustees took leave of their graves: "They ... kissed the earth. They had lived here for thousands of years ... this earth was made up of their ashes, of their sweat and blood. They kissed the soil, dug their nails into it, took handfuls and hid it in their clothing" (Kazantzakis 1974, 16). Here, a landscape is fully equated with communal identity.

The process of ethnic cleansing reached its low point in the 1930s and 1940s under Nazi Germany. The extermination of six million European Jews is one of the most devastating attempted genocides in history but it is often seen as different from other genocides because the Jews of Europe had no homeland at that time. However, they did have dwellings and often lived a "happy and secure middle-class existence [which was] interrupted by a lightning bolt of terror and followed by unspeakable agonies" (Deak 1989, 64). Deak also acknowledges that as the survivors become older, they can view their former "homeland" with some longing, despite the tragedies that occurred there.

On trying to re-establish their ancient homeland in Palestine, Jews were placed in the position of themselves perpetrating ethnic cleansing. The 1948 Israeli–Palestinian war is Israel's "War of Independence," but to the Arabs it is *al-Nakba* (the Catastrophe). It resulted in the expulsion of almost 800,000 Palestinians to refugee camps in adjacent countries (Har-Shefi 1980). Although Israel's official interpretation is that the Palestinians fled on Arab orders, both Israeli (Morris 1987) and Palestinian (Falah 1996) scholars attest to the existence of formal expulsion plans. Some of these plans – code-named "Broom" and "Passover Clean-up" – strongly suggest deliberate ethnic cleansing.

Yigal Allon, commander of the Jewish forces in the Galilee, specifically spoke of "cleansing" (Dalrymple 1997, 364).

Falah's (1996) work on Palestinian village annihilation, complete with six subcategories of destruction from "partial destruction" to "complete obliteration," suggests that of 863 Palestinian villages in what was to become the State of Israel, as many as 418 were totally depopulated. Of these, one-third suffered "complete destruction," with no walls left standing, while a further 19 per cent were utterly destroyed. In the case of "complete obliteration," village sites were levelled and physically annihilated by new forms of land use. Thus, Palestinian identity was obliterated from a "Judaicized" landscape cleansed of Palestinian markers.

Securing a salient from the Mediterranean coast to Jerusalem was a primary Israeli goal in 1948. This involved the destruction, on the western fringes of Jerusalem, of a series of Arab villages, the best known of which is Deir Yassin. Any captured Arab villages in the Jerusalem corridor were levelled to the ground. As Zablodorsky, Hagannah commander in Jerusalem, announced to Ben-Gurion: "The eviction of Arab Romema has eased the traffic situation" (Gilbert 1996, 188).

It was at this time that the Israelis began in earnest one of the cruellest punishments inflicted on a rebellious community: the dynamiting of family homes. At first, this policy was to encourage the Arabs to evacuate their villages. On arrival in Jerusalem, the Israelis used bulldozers to knock down and clear away Arab houses immediately to the west of the Wailing Wall. Later, the house-dynamiting policy was extended as punishment to the homes of "terrorists" or any house in which a Palestinian "freedom fighter" was thought to have spent even a single night. Within fifteen months of the beginning of the intifada in 1987, the Israeli army had destroyed or sealed nearly 200 Palestinian houses (Frankel 1989, 19). Although Yitzhak Rabin announced an end to this policy in 1992, his statement was immediately followed by the development of a new method of destroying homes using anti-tank missiles. In 2001, asking an Israeli soldier why her home was being destroyed, a Palestinian woman was told "revenge" (MacAskill 2001, 13). Several nations, included the United States, have condemned home destruction as contrary to international law, but the Israel High Court of Justice believes it is legal. Such practices only create more resistance. The army does not remove the rubble, but neither do the Palestinians, who contend that such sites serve as monuments to their cause.

The long road toward an Arab–Israeli peace settlement has also caused the evacuation of Israelis, both Jewish and Arab. Kliot (1983, 173–86) has provided a comprehensive analysis of loss of place associated with the need to resettle Israelis from the Sinai region as part of the Egypt–Israel Peace Treaty in the late 1970s. In this case, the Israeli settlement had special meaning because it represented ideological principles: settlements contributed to security of the boundaries of Israel; settling the land was a sacred mission; and the challenge to "make the desert bloom" was a mission of the state. It is therefore not surprising that the need to uproot these settlements was the cause for confusion and dismay. Displaced settlers referred to this action as a "holocaust" and to themselves as "refugees," "uprooted," and "evacuees." People immediately reacted with nostalgia, "walking to different spots and examining their feeling of attachment to those spots," or outrage: "Your shock is caused by the fact that your home is totally ruined. One moment you are with your roots deeply in the land and in another you are uprooted" (Kliot 1983, 173–86).

Another group of people was also affected by the displacement of the Israeli Jews from Sinai. When the Israelis were removed from Northern Sinai, there was a need to establish new military bases. Land for one of these bases was expropriated from an area inhabited by about 8,000 Negev Bedouin. There was no appeal against the *Negev Land Acquisition Bill* which became law in July 1980, and no negotiations were held with the Bedouin who had lived in this area for 130 years. According to Maddrell (1990, 11), when "a few [persons with political influence] saw the law as racist in victimizing a powerless minority," some land that was to be taken for the military base was returned for use by the Bedouin. However, compensation given the displaced Bedouin was only 2 to 15 per cent of that given the Sinai settlers. For many years, there were families who still lingered outside the air bases and others who would accept no compensation.

According to Israeli anthropologist Clinton Bailey (1994, A15), the roundup of the Negev Bedouin began as early as 1948 and was merely accelerated after the Sinai withdrawal. The Israeli government used the old argument that the resettled Bedouin would now have a better access to schools and medical care: "Dwelling in tents is uncomfortable and passé; life under a roof will bring them happiness and ease." As Bailey continues: "The cynicism is clear. No one asks whether Hasidic Jews are ready for the twenty-first century, so why ask it about the Bedouins? Jewish communities are not told to make way for the army, so why are the Bedouins? And why must 80,000 Bedouins, used to wide-open spaces, be limited to seven townships, when the 280,000 Jews in the Negev live in 114 villages and towns?" Many Bedouin regret their loss of the desert life (Meir 1997), and although they may have beautiful new houses, they continue to pitch their traditional black tents alongside them.

Perhaps the most controversial of all of Israel's colonizations is the one of the West Bank, an ancestral Palestinian population zone that was occupied by Israeli troops in 1967. In defiance of international law, Israelis have indulged in extensive geopiracy, forcibly taking over 50 per cent of Arab land in the occupied territories of the West Bank and Gaza (Said 1994, xxvii, 417) and creating therein extensive militarized zones and many towns and villages filled with hardline settlers. About 150 exclusively Jewish settlements have been built in occupied territory, with about 130,000 settlers in East Jerusalem and another 150,000 spread across land in what is one day to become the Palestinian Authority. To take a single example, the Palestinian village of Biddya, near Ramalah, is now overlooked by the fortified new town of Ariel, in which houses are available only to Jewish settlers. Beginning with only 8,000 Israelis, Ariel is projected to grow to ten times that number, which will mean the annexation of most of Biddya's agricultural land. Speaking of Ariel, a Palestinian remarked: "That was my grandfather's land. It has belonged to this village since the time of the Canaanites. But the Israelis took it in 1977. We've never received any compensation" (Dalrymple 1997, 345). Resistance by the inhabitants of Biddya has led to the destruction of village houses and the cutting down of its olive groves. Since most of the village income came from olive oil, its economy has been destroyed. In Arab Biddya, indoor plumbing is a luxury. In sharp contrast, Ariel's residents, some of them recent immigrants from Canada and the United States, bask in a US-style suburban environment with malls and swimming pools. After five years, such residents feel able to say: "It's home. It's impossible to put into words," while maintaining an armed militia that harasses Palestinians who have lived in the area for centuries (Friedmann 1989). When Dalrymple visited Ariel, a Canadian-Israeli, fearful that an Israeli-Palestinian peace accord could lead to the evacuation of her town, told him with no intended irony: "We're talking people's homes. You know what that means? *People's homes*" (Dalrymple 1997, 354).

Meanwhile, several million Palestinians languish in refugee camps in neighbouring countries. Even after fifty years, most have refused to be "resettled" permanently in Jordan or elsewhere. This refusal is keenly based on their sense of home, which, for Palestinians means an ancestral village in what is now Israel (Parmenter 1994). Several Arabic words can be translated as "home": while *dar* implies a structure, the commonly used *watan* refers to a land area or territory and has now come to mean homeland. It includes village, private crop and grazing land, common land, religious structures, and both natural and man-made landmarks. Women's traditional costumes vary from village to village, and Palestinian poetry and song often celebrates personal attachments to "Mother's house," "Father's orchard," and "beloved soil" (Taylor 1987).

If an Israeli – Palestinian Authority peace is ever finally implemented, we may look forward to yet another round of relocations. Bedouins may elect to reoccupy the desert, and thousands of Israeli Jews may be forced to leave the dwellings they have built in occupied Arab territory. Most difficult of all, refugee Palestinian Arabs, clutching copies of Walid Khalidi's *All That Remains* (1992), may try to return to the sites of their 418 lost villages. "It is a very sensitive issue ... this business of lost homes" (Grossman 1993, 37).

When Palestinians were relocated to the peripheries of their territory, in Gaza and the West Bank, they became "the nomad industrial reserve army in the socio-economic space of modern Israel" (Portugali 1989, 207). A similar situation existed in South Africa until the 1990s, whereby blacks removed to the peripheries of South Africa were used as a voteless labour force compelled to bus long distances every day to their work. Forced removals in South Africa, then, resulted in "forced busing" (Lelyveld 1985).

Although racial segregation had long existed in the country, the South African National Party's goal of social and political separation between whites and non-whites crystallized in 1948 into the policy known as "apartheid." Essentially, this policy involved: black development in separate states known as "Bantu homelands"; the establishment of white-owned industries on the edges of these Bantustans; and the clearance of blacks from "black spots" in designated white areas, with the relocation of these people to the Bantustans (Desmond 1971, 21). The ultimate result of this policy was to "purify" the South African state, with 3.5 million whites in control of 86 per cent of the land surface. Fifteen million blacks – 73 per cent of the country's population – would have to crowd onto a mere 13 per cent of its land. The policy was extended to Namibia in 1967. Dugard sees this project as the attempted implementation of white fantasies: "Pretoria has set in motion the implementation of its ultimate fantasy – a South Africa in which there are no black South African nationals or citizens; a South Africa that cannot be accused of denying civil political rights to its black nationals for the simple reason that there will be no black South Africans, only millions of migrant workers (or guest workers, as the fantasy sees them) linked by nationality to a collection of unrecognized, economically dependent mini-states on the periphery of South Africa" (in Platsky and Walker 1985, 16).

To deal with racially mixed urban areas, the *Group Areas Act* was promulgated in 1950. "An instrument for institutionalizing the disadvantage of those not in power" (Western 1981, 234), the act not only required that millions of blacks be uprooted from the cities and decanted into dubious homelands, but also provided for the setting up of separate areas in every city and town for each of the racial groups identified in the *Population Registration Act*. Besides the usual categories of white, black, coloured, and Asian (mostly Indians), an attempt was also made to concentrate the scattered Chinese as well as the Cape Malays. The creation of this "tidy racial geography" (Smith 1978, 87) had by 1975 resulted in the resettlement of 305,739 coloureds, 153,756 Asians, and 5,898 whites. The ultimate aim was to relocate one in six of the coloured population, one in 3.5 of the Indians, and one in 642 of the whites. By 1984, this ludicrous game of musical chairs had succeeded in displacing over 83,000 coloured, 40,000 Asian, and as many as 2,428 white families (Rogerson and Parnell 1989, Christopher 1991).

But the relocation of blacks was on a much larger scale altogether; over one million had been moved as early as 1975 (Smith 1978, 87). They proved easier to remove, however, as they had been segregated since the mid-nineteenth century, were not entitled to own land outside proclaimed rural reserves, and were legally regarded only as "temporary sojourners" in South African towns (Christopher 1991, 241). Further, they had no vote, were not consulted, and most had little understanding of what was happening to them other than that their land was needed by the government (Platzsky and Walker 1985, 46).

Government rhetoric sought to soften the blow. When the word "apartheid" became unacceptable, it was replaced by the less offensive "separate development," and later with "multinational development" and "plural democracy." Many hundreds of thousands were moved into new villages under the rubric of "betterment planning." In relation to the cleansing of blacks from towns, J.J.G. Wenzel, deputy minister of Development and Land Affairs, told the Drifontein Community Board in 1981: "The removal and relocation of so-called black spots is carried out in accordance with a policy which has as its goal the improvement of the standard of life of all people of South Africa. You will therefore appreciate that it sometimes becomes necessary for people to be encouraged to move for their own good" (Platzky and Walker 1985, 171).

Desmond (1971, 3), who saw the population of the "black spots" in Dundee, South Africa removed to an area called Limehill, described a pattern that was to become familiar. An official of the Department of Bantu Administration and Development (with the unfortunate acronym BAD) arrived in 1965 to tell the tenants of the South African

Mission that they were to be moved the following summer. In 1967, another man came and painted numbers on the doors of the black homes in the Mission. Later that year, the Bantu commissioner himself arrived to tell people that they would be moved in May 1968 but that they would have the opportunity to build new houses and schools and that other facilities would be ready. Finally, a letter announced that the Mission was to be cleared on January 20, 1968.

Unsuccessful attempts at resistance were made by nearby white residents, including representatives of many churches, but the black Africans were given no say in their future. Their reaction was: "We are suffering. We have been thrown away. We have nothing. But what can we do?" (Desmond 1971, 7). Apparently with little protest, people were removed to Limehill where they found nothing but the bare *veld* (open country) and a pile of folded tents. When questioned in the House of Assembly, the minister of BAD said that the removals were voluntary and had been effected humanely.

To forcibly relocate the inhabitants of major cities was a formidable task. In this connection, the experience of Cape Town has been extensively studied. The least-segregated South African city in 1950, Cape Town had become by 1985 the most racially divided because of the severity with which the *Group Areas Act* was applied. One of the goals was to preserve the city's central area as a symbol of white history, although a small Malay quarter was also preserved. It was particularly important to rid the inner city of mixed-race neighbourhoods. Thus, the square kilometre of District Six was levelled to the ground and all the coloured, Malay and Indian residents of the Loader area were expelled. This expulsion was particularly bitter for the coloured people, who may not have taken a special pride in being coloured, but who certainly took pride in being from Cape Town: "*Place* of origin – home – has become an essential element of self-definition for Coloured people" (Western 1981, 149).

In a detailed study, Western (1981, 218ff) interviewed many of the former residents of the district of Mowbray who lost their homes. One old man, a veteran of two wars, said: "You make a place of your own, you make it comfortable for your old age, then they come and tell you you've got to go. And you can't start again, time's against you – you remember those people who committed suicide in Tramways Road, Sea Point? They gave me a month's grace to build on my son-in-law's land … I don't think they gave us a true value for our house." And another commented: "A lot of people died after they left Mowbray. It was heartbreaking for the old people. My husband was poorly, and he used to just sit and look out the window. Then before he died, he said, 'You must dress me and take me to Mowbray. My mum and dad are looking for me, and they can't find me [except in] Mowbray.' Yes, a lot of old people died of broken hearts."

Squatter settlements were also targeted. In 1983, the building of the new town of Khayelitsha on the outskirts of Cape Town was announced. About 100,000 local blacks with rights to live in the western Cape would be relocated there, while another 100,000 "illegal" blacks would be "repatriated" to distant homelands. The aim was to cleanse the Cape peninsula of black squatter settlements, perhaps the best known of which is Crossroads: "During the period 17 May to 12 June 1986, Cape Town and the international community witnessed the most brutal destruction and forced removal of squatter communities in this country's history. In a period of less than four weeks, an estimated 70,000 squatters ... became refugees in their own land" (Cole 1987, 131). One year later, on 25 June 1987, this author (Porteous) witnessed a subsequent raid on the remains of the settlement of KTC, a terrifying affair involving thirty armoured troop carriers full of soldiers in combat gear, with the operation directed from a helicopter clattering overhead.

In 1991, the *Land Act*, which had reserved 86 per cent of the land for whites, was repealed. In order to provide an orderly change to a new system, communal tenure was proposed, but this system ignores the importance of land to the African people (Mallaby 1992). It also ignored the massive post-1993 demands for a return to former homes and lands, legacy of "the massive imbalance of power ... between those who remove and those who are removed" (Platzsky and Walker 1985, 176). Afrikaaners may now face domicide in their turn. The democratization of the Republic of South Africa in 1993 has inevitably led to problems as the "homelands" lose their status and people try to return to their former homes.

Similar returns are being attempted in ethnically cleansed areas of the 1990s in both the former Yugoslavia and the countries of Rwanda and Burundi. "Ethnic cleansing" became a common phrase during the war between Serbs, Croats, and Muslim Bosnians which killed as many as 200,000 people between 1991 and 1995 and later when much of the ethnic Albanian population of Kosovo was driven out by Yugoslavia in 1998–9. The atrocities that took place then were, in fact, merely a revival of activities that took place during the Second World War, when the pro-Nazi Ustache party stated that of the 600,000 Serbs living in Croatia, a third would have to convert from Orthodoxy to Catholicism, "a third expatiate themselves, and a third die" (Hartmann 1992, 18). The ethnic cleansing of the 1990s is said to be the idea of sixteen members of the Belgrade Academy of Sciences and Arts who circulated a secret memorandum in 1986 calling for

proper treatment for the Serb nation and the Serbianization of territories to which they laid claim. These policies have awakened long-
standing desires for revenge.

Resulting from these hostilities, the destruction of the cities of Sara-
jevo and Mostar caused an outpouring of Bosnian emotion at the
deliberate annihilation of national libraries, museums, cemeteries,
mosques, and archives: "The history of our homeland is gone" (*Manchester Guardian Weekly* 1992, 7). Here the destruction becomes
memoricide and the losses are irreparable: the university's holdings; the
national archive of newspapers and periodicals; a collection of oriental
manuscripts; the historical archives of Hercegovina before the
Ottoman conquered the region in the fifteenth century; a library of cal-
ligraphic and illuminated manuscripts dating from the twelfth century;
and, ironically, an anthropological collection that included records of
Serbian civilization. The toll on the human population was indeed
heavy. As old people die or are killed, and the physical structures and
records of villages are destroyed, the very remembrance of these villages and their homes is expunged (Wilkes 1992, 22–25).

But memory persists. White racist mobs burned down the Tulsa,
Oklahoma black neighbourhood of Greenwood in 1921 and completely destroyed the small black township of Rosewood, Florida, in
1923, forcing residents to relocate. Not until sixty-one years later did
the state of Florida compensate Rosewood's survivors, while the fight
for compensation in the Oklahoma legislature was still going on in
2000 (*Washington Post* 1993, Calhoun 1997, Yardley 2000). Domicide is not difficult; memoricide proves a much more arduous task.

Settlement Rationalization

Ethnic cleansing generally involves a contest between groups over
which will occupy a particular place, having "purified" it of the other.
Settlement rationalization is frequently a contest between administrators who wish to abandon certain settlements and regroup their former
inhabitants according to a new master plan and those residents who
would prefer to remain in their time-honoured communities. The
implementation of political theory or the search for administrative convenience are motives for settlement rationalization.

Two cases from adjacent countries, Mozambique and Rhodesia/
Zimbabwe, point out the differences in motive but similarities in result,
which characterize ethnic cleansing and settlement rationalization. In
Rhodesia, the 1931 *Land Apportionment Act* granted twenty million
hectares of land to 50,000 white persons, while one million black
Africans were confined to nine million hectares. This act was also to

serve as the main vehicle for segregation policies for the next forty-six years (Meredith 1979, 21). As a result, there was a massive relocation of black people who had both spiritual and economic ties to their land (Sylvester 1991, 35). In 1982, faced with the results of a major drought, resettlement was again instigated as the new country of Zimbabwe strove to deal with victims of past land appropriation policies by redistributing land and improving rural conditions. By 1987, 35,000 people were again resettled, perhaps foreshadowing events to occur later in South Africa.

In contrast, in Mozambique, the new social and economic order embracing Marxist-Leninist principles brought in by the Frelimo Party (Frente Libertação de Moçambique – the Mozambique Liberation Front) in 1977, was based on the creation of communal villages, state farms, and collectives. A similar system of *Ujamaa* cooperative villages had been instituted earlier in Tanzania. Much of Mozambique's rural population, estimated at ten million persons, had lived in individual homesteads or small villages. The communal villages were intended to provide for a better level of services such as health and education as well as a nexus for the new social and political order. Much success has been attributed to the formation of these villages in terms of the population attracted, the new participation by women in the political processes, and the ending of certain social practices such as initiation rites, child marriage, polygamy, and marriage payments.

However, Hanlon (1984, 129) provides an example of villagers who, in 1981, were given just two weeks to move from their homes into the new village of Garuzo. While the Frelimo Party was said to understand that people do not easily give up their old homes and that in such cases they should be allowed to remain, local administrators frequently forced people to leave their homes. In addition, the process was slow since little assistance was given to the program. Of more than 1,000 villages built since independence in 1975, only 200 reached a full stage of development in the first eight years. Most villages were created in places in which war or natural disasters had destroyed homes or to which people were otherwise forced to move. As Job Chambal, National Director of Communal Villages, stated: "When it is said that we are forcing people into communal villages, it is true. Because if we don't, then the enemy will use these people to destroy their own future. These people are being liberated" (Hanlon 1984, 128). In fact, Mozambique's post-independence problems were worse than the challenge of getting rid of the Portuguese (Harding 1993, 14). The apathy of the peasant population regarding the Frelimo reforms played into the hands of the rebel movement, the Renamo, which was initially formed by Rhodesian officers and also disaffected Mozambican guerrillas. Indeed, the villagization plan proved so disastrous that it had been largely abandoned when this author (Porteous) visited Mozambique in 1989. It has been described as *"une violence culturelle"* (Newitt 1995, 549).

Political theory, of course, has been the background to numerous population shifts in the socialist world, the most horrendous being the Khmer Rouge's evacuation of all Cambodian cities in Year Zero (1975), their inhabitants then being worked to death in rural killing fields. A more accessible case studied by Porteous (1972) concerns the desire of Chilean governments in the early 1970s – both Christian Democrat under Eduardo Frei and Socialist under Salvador Allende – to lessen their country's symbolic dependence on the United States. In 1970, having nationalized the copper industry, then Chile's greatest export earner, the next goal became to reintegrate the Chilean populations of American-built company towns at mine sites into the mainstream of Chilean life. Although some residents were indeed moved, the coup by General Pinochet in 1973 put paid to these politically motivated relocations.

Equally controversial and also ending in a coup were the grandiose projects of Romania in the 1980s. While Romania's leader at the time, Nicolae Ceausescu, may have begun with a desire to improve social conditions in his country, this desire was supplanted by ideology and megalomania. A document from the twelfth congress of the Romanian Communist Party speaks of progress and civilization, and the creation of "the new man" through "the new humanism." It declared that ethnic differences would be resolved by "liquidating inequalities and discriminations," called for "increasing homogenization," and mentioned that one means of achieving these goals would be a "program of urban and rural systematization" (Malcolmson 1994, 5).

What systematization meant in practice during the 1980s was the destruction of forty towns, the historic buildings of which were levelled. Ceausescu's next plan was to raze one-half of Romania's villages by the year 2000 and eliminate smaller communities as rapidly as possible. The justification for this action was the need to gain more arable land and make modern facilities such as schools available to rural communities. The Hungarian and German ethnic communities were most severely affected before Ceausescu's reign was halted. One example of this destruction is the following: "Vladiceasca, once a thriving little rural community of about 80 houses, vanished in the summer of 1987. It fell victim to Nicolaie Ceausescu's manic dream of 'systemization,' his plan to wipe out thousands of Romanian villages and move the inhabitants to concrete blocks where they would be more controllable ... Most peasants were already so intimidated by decades of oppression that they obliged ... [one of these] Mr. Nastase says, 'I would have become 100 years old in my own house.' He shows a small painting of it. 'Now I just want to die'" (Elmendorp 1990). While this resettlement began with a motive of socio-economic improvement, the real motive behind the destruction of Vladiceasca is seen to be linked to a totalitarian ideology. When Communist Party members sought to restrain Ceausescu's megalomania in a letter of March 1989, they stated their complaint as follows: "Romania is and remains a European country ... You have begun to change the geography of the rural areas, but you cannot move Romania into Africa" (Malcolmson 1994, 59). Ceausescu was arrested and then executed on Christmas Day, 1989, but not before he had created chaos and misery through thousands of forced removals.

A somewhat more benign ideology of administrative, social, and political convenience led to a spate of settlement rationalization in subpolar regions after the Second World War. This involved the abandonment of small outports and the regrouping of their inhabitants in larger, more accessible towns in Norway, Canada, and Greenland. In the 1940s, one-half of Newfoundland's rapidly growing population lived in 1,200 settlements, each of less than 500 people, scattered along a lengthy coastline. Provincial governments attempted to find solutions to the social and economic problems of such communities by initiating a move away from primary industries, such as the declining inshore fishery, while simultaneously regrouping existing industries. As early as 1954, with the first Centralization Program, 115 communities were evacuated, totalling over 1,500 families and 7,500 persons. In 1965, the Canadian federal government began to play a major role, and the Federal–Provincial Newfoundland Resettlement Program was launched via the *Resettlement Act*. Under this act, a further 137 isolated outports scattered along the coast of Newfoundland were abandoned by 1972. Like all government programs, this one developed a language of its own in which people were moved from "designated outports" to "approved land assembly areas" or "major fishery growth areas." Copes (1972, 128) acknowledges that "many resettlement officials, under the influence of their own commitment to, and enthusiasm for, resettlement, used their powers of persuasion to convince many outporters they ought to move" and then a "moving fever" took over as those who were reluctant to leave did not want to be left behind.

For the residents of the outports, the future was uncertain whether they moved or stayed. Were they to stay, their small communities would provide but a few months of hard labour at fishing followed by a long period of unemployment and welfare. Yet, should they move to a more centralized community, they would not have the job skills to compete in the urban environment (Matthews 1976, 2). The National Film Board of Canada produced a bittersweet memory of outport life in the short film *Children of Fogo Island* (1962), which captured forever the activities of children – a ramshackle shed as a playhouse, stilts, and small wooden boats – all against a harsh backdrop of sea and rocks. Yet, somehow it appears an idyllic place for children to call home, and, in fact, many residents have clung to "the Rock" (*Maclean's* 1993, 22).

Gordon Pinsent, the noted actor and writer from Newfoundland, demonstrated the pain associated with being forced to leave one of these communities in *John and the Missus* (1977), a film that he wrote, directed, and starred in. Clearly depicted were the slick government agents who encouraged people to leave, buying off some in order that others would follow, threatening the laggards with being left behind. To quote the slightly mad postmaster who found acceptance in this community: "They're dancin' with all of us." The character John, played by Pinsent, eloquently sums up the plight of the resettled: "Pack up the town – home – next thing I know they'll say there was no one here at all – no names. How do you resettle? It might be all right for you fellows, but we made up our minds ... we're all ready settled ... Go back and say you can't find us. We were too small to see with the naked eye. Who am I going to know alongside me when they put me there? Not a blessed one. You're telling me we're going to die."

General information on the reaction of persons who lost their homes or communities in Newfoundland is available from a survey conducted by Matthews (1970, 311). He found that 50 per cent of those he interviewed wanted to move, while another 33 per cent felt they had no choice. According to Copes (1972, 123), older people found it difficult to move but had little choice when the whole settlement was leaving. The new locations were not without problems. There was often significant unemployment, and a resettlement welfare ghetto emerged in the new location. The resettled frequently suffered a loss of self-esteem, and there was considerable deterioration of the social environment.

Matthews also examined the communities of Small Harbour, Mountain Cove, and Grand Terre which the inhabitants refused to leave. In the case of Small Harbour, the community inadvertently found out that it was destined for resettlement, which enabled community leaders to organize against government policy. In the case of Mountain Cove, the community simply ignored the threat of being moved as they saw no reason to leave. Grand Terre survived due to the efforts of one woman who challenged rumours that a mass exodus was expected from the community. The resistance of its residents is shown in the following quotations: "The way it is here, they're now putting concrete pillars under their houses. They're going to stay here until they got no other remedy," and "Right now, I'm so far back in age, in education, so I might as well stay with what I got here ... I've got a half decent home" (Matthews 1976, 43, 102).

There are differing views about the success of these resettlement programs, and hindsight provides an even crueller judge when seen in the light of the serious problems of Newfoundland's fishing industry today. Copes (1972, 107) suggests that the programs are "too readily seen as a bribe to induce households to move against their better interests and instincts" rather than in the context of the massive assistance given to inshore fisheries, which alone made life supportable in the outports. He suggests that simply closing off these services would have been inhumane and impractical and that people would not have been able to move themselves. Further, he believes that a "welfare mentality" existed already, and that most fishermen wanted improved schooling and medical facilities and a better future for their children. He quotes two surveys in which 62 per cent of those surveyed indicated a willingness to move. In sharp contrast, Goulding (1982) identifies the Canadian state, acting in the name of corporate capitalism, as responsible for the crisis in Newfoundland. He contends that the state gives huge grants of money and land to major capitalist investors, creates laws that prevent others from using this property, and moves people from fishing communities to supply cheap labour for heavy industry. In the eyes of the then premier of Newfoundland, Joey Smallwood, this was progress.

Unfortunately, with the collapse of the Atlantic cod fishery in the early 1990s, many of the remaining Newfoundland villages were again threatened with extinction. By 1993, there were new calls for resettlement, led by the economist Parzival Copes, who supported the earlier resettlement program (Bailey 1993). But resettlement remains "a four-letter word" to many Newfoundlanders, who believe that another top-down settlement rationalization program is both economically infeasible and socially undesirable.

Similar beliefs that a better living could be provided elsewhere caused the uprooting of some native Inuit persons in Northern Canada. In August 1953, the Canadian federal government forced seven families from Inukjuk, Northern Quebec, and three from Pond Inlet on the northern tip of Baffin Island (a total of eighty-seven people), to move to Grise Fiord on the southern tip of Ellesmere Island. This action, described as an "experiment" in one 1953 memorandum, was justified by the federal government on the basis of poor hunting in northern Quebec and the belief that the Inuit should be assisted to return to their original lifestyles.

In sharp contrast, the Inuit believed that the federal government was moving them for political reasons so that their new settlements would assert Canadian sovereignty over the Arctic Archipelago in the face of American pressure to have the region declared international waters. As a result of the move, these people were exposed to horrific conditions in their first year: "When we finally landed [at Grise Fiord], it was as if we had landed on the moon, it was so bare and desolate," says one now-elderly exile (Ackerly 1993). "There was no food and no shelter" (Williams 1993, 85). Other survivors speak of relatives who "died of a broken heart" within a year of arrival.

In their book *Tammarniit (Mistakes)*, Tester and Kulchyski have provided a fascinating analysis of the Inuit's relocation in which they see the state as a totalizing influence and suggest that "much of what is observed ... closely parallels present day attempts to bring indigenous and local cultures around the world into a web of international capitalist relations" (1994, 5). They point to the fact that the Canadian government was entering a period of welfare state reform in 1953, and this converged with the state's concern for territorial integrity during the first wave of post-war government expansion in the north. Yet, the "High Arctic Exiles," as they are now called, somehow managed to survive. When pressured in 1989, the government paid for the return of forty Inuit people to their former homes, causing a breakup of families along generational lines. For the Inuit, the High Arctic settlements have become "an icon of pain, endurance, and betrayal," but those who stay, stay by choice and are "fiercely committed" to their new home (Williams 1993, 85).

In 1994, a royal commission set up to enquire into the affair, reported its findings. Throughout the controversy, conflicting evidence was given. While the Inuit continued to claim that they had been forcibly moved, officials countered that the Inuit moved voluntarily (Ackerley 1993). Bent Sivertz, who was in charge of the 1953 relocation, suggested that many witnesses had refashioned their stories in order to claim compensation, and he believed the relocation to be "a heart-warming success" (Murphy 1995, A14). Nevertheless, the Royal Commission on Aboriginal Peoples report recommended that the Inuit should be compensated for their pain and suffering, and a $10 million High Arctic Trust was set up to provide economic, cultural, social, and educational benefits to the surviving relocatees and their descendants.

Like the Canadian Inuit, their counterparts in Greenland across the Davis Strait were subjected to insensitive relocation organized by whites from faraway cities. After the Second World War, the Danish government decided that if its colony was to be economically viable and its people to enjoy a full range of modern services, the Inuit would have to abandon their tiny isolated settlements, and the population had to be centralized. All the planning was done by "well-meaning but arrogant" Danes (Hall 1987, 127), and the victims were not consulted. In a massive development program, scattered Inuit hunters were moved to four towns of between 3,000 and 4,000 people located on the ice-free southwest coast of Greenland. Fishing was encouraged here, but in the hinterland, investment was withdrawn from "unprofitable" villages and schools and stores were closed.

Planning long-distance from Copenhagen, the Danes completely changed the character of Nuuk (formerly Godthab), the Greenland capital, erecting massive six-storey apartment blocks into which the former nomadic hunters were crowded. This ludicrous plan demonstrates just how far the governing whites were from understanding the Inuit and how determined they were to modernize their colony come what may. As might have been expected, the imprisoned Inuit degenerated through psychological problems, despair, broken families, alcoholism, crime, and suicide.

Nuttal lived with the Greenland Inuit and, as a prelude to his discussion of kinship and community, explains how, as in Canada, these people had initially been encouraged to gather into their "unprofitable" isolated settlements by missionaries and traders. The subsequent move to the towns typically caused "fragmentization of the kin-based groups that characterized village life. Incomers to the towns suffered from economic isolation, marginality, and discrimination" (1992, 19). Ethnic conflict with Danish transient workers became common.

The coming of Home Rule in 1979 encouraged the creation of a national character, but even in the 1990s, urban Inuit still identified themselves with special localities elsewhere, the "last outposts of real Greenlandic culture" (Nuttal 1992, 21). Unfortunately, further centralization is planned, ringing the death knell of many of the remaining settlements of the north and east of the island. Further, European anti-sealing campaigns have destroyed the economic base of many Aboriginal hunters in polar regions. "The seal hunter is a disappearing species" (Hall 1987, 137).

This whole sorry episode of domicide and cultural genocide is summed up by Philip Lauritzen, a Danish spokesman for the Home Rule government: "I think we could experience a very sad thing here in twenty years, when European anthropologists come up here to interview hunters in apartment blocks, wanting to know why they left their settlements. Then, everyone in Europe will ask, 'who could ever destroy that kind of culture?' That, sadly, is actually the situation that

Greenpeace, and the World Wildlife Fund, and many others have created" (Hall 1987, 137).

CONCLUSION

This extremely wide-ranging discussion of extreme domicide via war, colonization, and resettlement suggests some tentative conclusions, which will be fleshed out in chapter 6. First, domicide is usually planned by powerful leaders – whether war chiefs, top-level bureaucrats, or even intellectuals such as those that generated the idea of apartheid or the proposal for the ethnic cleansing of greater Serbia. Only occasionally does the concept become one of mass or sectoral popularity, as with white settlers during the geopiracy of indigenous lands in North America. Second, domicidal plans are often enshrined in, and legitimized by, legalities, manifested in bureaucratic decrees such as a resettlement act or the *Negev Land Acquisition Bill*. Third, those who actually carry out the most violent domicidal plans, having no good choices, turn to national interest and self-preservation to justify their actions. Ultimate motives vary from vengeance and profit to administrative tidymindedness and "efficiency."

Those responsible for domicide do not communicate effectively with those who are to be moved. Because the latter are not included in the decision (hardly possible, of course, in war), they inevitably become victims whose loss of home, at whatever spatial level, is likely to result in stress and trauma. As Colson has pointed out, some people may welcome change even in extreme circumstances, but most "probably like variety only so long as it is an embroidery upon the reassuring familiarity of customary routines, well known paths and scenes, and the ease of accustomed relationships" (1971, 1). As we have shown above, however, accustomed relationships, even those of the family, readily break down with the experience of extreme domicide. The relocated are frequently disempowered and suffer socially and economically.

It remains to be seen whether any of these characteristics of domicide are apparent when we move from war, geopiracy, and major resettlement projects to the everyday loss of home through urban development and the building of public infrastructures. Paradoxically, extreme domicide appears to be both avoidable and reversible, whereas with everyday domicide, homes may be lost forever under the grinding juggernaut of "progress."

CHAPTER FOUR

Everyday Domicide:
Landscapes of Cruelty

The art of losing isn't hard to master;
so many things seem filled with the intent
to be lost that their loss is no disaster ...

I lost two cities, lovely ones. And, vaster,
some realms I owned, two rivers, a continent.
I missed them, but it wasn't a disaster.

– Even losing you (the joking voice, a gesture
I love) I shan't have lied. It's evident
the art of losing's not too hard to master
though it may look like (Write it!) like disaster.
Elizabeth Bishop, "One Art,"
Geography III, 1976

Extreme domicide tends to happen infrequently and usually to people
unlikely to be reading this book. In contrast, everyday domicide occurs
continuously all over the world and can affect everyone except the
wealthy and those who are its perpetrators. Unlike extreme domicide,
the everyday variety comes about because of the normal, mundane
operations of the world's political economy. It is brought about, first, by
inequalities based on the division of the world into rich and poor, colonizer and colonized, city and countryside – factors recognized alike by
the fourteenth-century Islamic writer Ibn Khaldun, the fifteenth-centu-
ry Florentine Niccolo Machiavelli, the nineteenth-century's Karl Marx,
and the early twentieth-century's Vladimir Ilich Lenin. It was the latter
who so rightly endorsed Hilferding's assertion that "Finance capital
does not want liberty, it wants domination" (Lenin 1977, 32). Anyone
living in the capitalist world – which is, for all intents and purposes, the
whole world at the turn of the twentieth century – is aware that large
corporations, transnationals, and banks can alter landscapes and lives
almost at will. This is a world of power that brooks little opposition.

Domicide is brought about, secondly, by the willing co-operation of the majority of the populations of industrialized states and of the elites in Third World countries. Nations composed less of citizens than of consumers (the word has an ugly edge to it) rapidly turn luxuries into necessities and wants into needs. For the consuming desires of the urban masses, airports and highways must be built, city centres redeveloped, and resource extraction extended yet further into remote hinterlands. Urban Disneylands are matched by wilderness national parks.

Encouraged by bureaucratic thinking, dwellings are regarded as "shelter" or "housing." Encouraged by capitalist thinking, dwellings become commodities to be traded. In an increasingly mobile world, loyalties to neighbourhoods, regions, or even countries are discarded. In a world in which no one invested love or memories in physical structures or landscapes, and in which compensation for relocation was adequate, domicide would no longer occur. Happily, the power economics of globalization have not yet reduced all of us to the status of rootless consumers, eager only for the next technological fix. Loyalty to place remains strong, and while it persists, forced relocations for power, profit, and pleasure will continue to involve domicide and, one hopes, resistance. In a less pitiless world, perhaps, the art of losing would not need to be mastered.

URBAN AND ECONOMIC DEVELOPMENT

Never before have so many people been uprooted. Industrialization ... requires ... a new kind of violence.

John Berger (1984, 55)

The Industrial Revolution began in Britain in the late eighteenth century and spawned mass urbanism in the following century. From the late nineteenth century, the wholesale relocation of populations from countryside to city has proceeded at an unparalleled rate, first in the industrialized world, and after the Second World War, in the Third World. Students of urbanism are often able to apply their first-hand experience or that of their recent forebears to this latest of several fundamental changes in the human condition.

Whereas in the early 1970s only one continuously urbanized area exceeded fifteen million inhabitants (New York / Northeast New Jersey), a number of cities had reached this level by 2000, and cities of one million are now commonplace. More significantly, the proportion of the world's population living in urban environments has increased rapidly, and early in the twenty-first century, urban living will be the normal way of life for a majority of the population of the globe.

The problems of cities are legion and have generated an enormous literature in the social and planning sciences. From the point of view of potential domicide, the relentless growth of cities outward across agricultural land and upward in their centres in a frenzy of redevelopment is cause for concern.

Urban Redevelopment

As one millennium gives way to another, half the world's population inhabits urban environments. During the twentieth century, the transition from rural to urban to suburban life has been rapid. Industrial countries reached maximum urban populations of 70 to 90 per cent of the total often before 1940; Third World nations began to follow suit after 1950. Today, cities are "property machines" (Ambrose and Colenutt 1975); each is a "Manipulated City" (Gale and Moore 1975) subject to the operations of powerful economic and political institutions.

It was such a combination that produced the American federal program known as "urban renewal," which began in 1949 and continued into the late 1960s. This program provided city renewal agencies with federal funds and the powers of eminent domain "to condemn slum neighborhoods, tear down the buildings, and resell the cleared land to private developers at a reduced price. In addition to relocating the slum dwellers in 'decent, safe, and sanitary housing,' the program was intended to stimulate large-scale private rebuilding, add new tax revenues to the dwindling coffers of the cities, revitalize their downtown areas, and halt the exodus of middle-class whites to the suburbs" (Gans 1965, 29). Similar projects, often more honestly called "slum clearance," were in operation in Britain and across Europe. Soviet-influenced nations also pursued this course, with appropriate modifications. The result was that by the 1970s, it became, at first sight, difficult to tell the difference between the low-income tower block regions at the edges of cities in North America, Europe, or the USSR.

It is generally agreed that the results of these programs were equivocal at best, and that socially, a great deal of individual and communal misery was caused (Porteous 1977). In the United States, it was found that many renewal sites were chosen not because they were the slums most in need of renewal, but because they offered the best sites for luxury housing. Indeed, far from helping the poor to rehouse on-site, projects typically provided cleared inner-city land for the expansion of government offices, public institutions and even shopping areas. A

1961 study of renewal projects in forty-one American cities found that 60 per cent of dispossessed tenants were merely relocated to other slum areas (Gans 1965, 30). Concern for uplifting the conditions of the poor was merely a smokescreen to produce cheap land for the property machine; between 1949 and 1964, the proportion of federal urban renewal expenditures devoted to the relocation of families and individuals was only 0.5 per cent.

Because the chief effect of urban renewal was to destroy low-rent housing, and because renewal pulled down more houses than it built, it was strongly criticized (Jacobs 1961, Wallace 1968, Sayegh 1972, Porteous 1977). In sum, by attacking the physical expressions of poverty, notably substandard housing, *The Federal Bulldozer* (Anderson 1964) of urban renewal became a vicious tool dedicated to the perpetuation of poverty.

Although "urban renewal" and "slum clearance" projects declined remarkably in the 1970s, large-scale displacements continue in many cities. Canary Wharf in London is a recent example. Further, the process of gentrification, whereby the middle classes usurp the inner-city neighbourhoods of low-income people, leads if not to eviction, then to "encouraged" resettlement (Smith 1996). Rohe and Mouw (1991, 57) note that in 1987 alone, 12,000 households were directly displaced by Housing and Urban Development and Department of Transport projects in the United States. They suggest that while the economic impacts of such displacement are now better mitigated, serious social impacts still continue. It is to these that we now turn.

The development programs of several decades ago are of interest here because they stimulated the first detailed studies of domicide. By 1960, urban renewal projects in 200 American cities had displaced about 85,000 families, their relocation having "provided only marginally better housing, in very similar neighbourhoods, at higher rents," and having "done as much to worsen as to solve the social problems of the families displaced" (Marris 1969, 123).

The most thoroughly studied project, from the point of view of domicide, was Boston's West End. A thriving multi-ethnic inner-city neighbourhood – mainly Italian, Jewish, and Polish – the area occupied forty-eight acres of the seventy-two-acre working-class district of downtown Boston. Mainly consisting of five-storey walk-up apartment blocks, the area housed 12,000 residents in 1950 and still had 7,500 (in about 2,800 households) when the city took the land for redevelopment via eminent domain in May 1958. With a lively street life, local traditions, and a high level of both tolerance and informal social control, the area did not "satisfy the social criteria which would make it a slum" (Gans 1972, 193).

We are fortunate that a number of ethnographers, sociologists, health workers, and persons working in the realm of community studies chose to perform detailed research on the effects of the forced relocation on the West Enders. Ryan (1969), for example, discovered that the West End had a distinctive local culture. Whereas occupation is important in the middle-class configuration of identity, in the West End, the most salient feature was friendship – and local friendship at that. Ninety per cent of the residents identified themselves specifically as West Enders, perceived a sharp boundary between the West End and the rest of Boston, and formed a distinct socio-cultural system. To "marry outside" the area, for example, was to risk losing the benefits of this neighbourhood-level home. A battery of psychological tests indicated that West Enders differed in several ways from the dominant contemporary American pattern. In particular, they emphasized "integration" and "ethnic harmony" as important values, while their "expressiveness" involved setting a high value on human relationships, including "helping others." Personal striving toward social power, wealth, and high-status occupations was acknowledged, but with the rider that the achievement of such goals would involve the sacrifice of personal contentment and social relationships. In other words, the dominant, striving society destroyed the West End neighbourhood, which had a different non-striving but vibrant, functional subculture with a strong geographical base – something we soon came to call an "urban village" (Gans 1962).

The loss of such close-knit ties upon redevelopment and forced removal is likely to cause both social disruption and personal trauma. Residents could not be moved *en bloc*, nor could their particular urban environment be reproduced. Moreover, they could not stay in place because the area – adjacent to downtown, the high-income area of Beacon Hill, and the Charles River – was a prime target for replacement by a high-rent complex of offices, government centres, luxury apartments, and extensions of the Massachusetts General Hospital. Only 15 per cent of the displaced found replacement housing through the aid of renewal officials. Most had to disperse to other areas of Boston with the same high density and mixed land use environment as the West End (Hartman 1966). Such replacements of so-called slum housing have been condemned as mere slum relocation, or as "land grabs aided by government subsidies and the powerful privilege of eminent domain" (Higbee 1960, 86). Among blacks, urban renewal was wryly renamed "negro removal."

Above all, forced relocation results in identifiable social and personal effects. Several reports on the West End agree that many of its inhabitants had strong attachments to the area and did not wish to leave.

Forced relocation from this supportive environment proved disrupting and disturbing for many. The loss of close spatial links with friends and relatives and the loss of a feeling of enclosure and safety were very apparent. Hartman (1966) found that 76 per cent of West Enders had unreservedly positive feelings about the area and reported potential relocatees as saying: "I love it. I was born and brought up here. I like the conveniences, the people, I feel safe ... I'm going to miss it terribly" and "I loved it very much. It was home to me. I was very happy. Everyone was very nice. All my relatives lived there." Gans (1962, 43) also described the feelings of persons about to be relocated: "It isn't right to scatter the community to the four winds. It pulls the heart out of a guy to lose all his friends" and "I wish the world would end tonight. I'm going to be lost without the West End. Where the hell can I go?" In what became a classic study, Marc Fried first indicated that many sources of satisfaction could be obtained in an "urban slum" (Fried and Gleicher 1961) and then sought to tap the feelings of people who had been forced to move. In "Grieving for a Lost Home," Fried (1966) asked his respondents questions such as: "How did you feel when you saw or heard that the building you had lived in was torn down?" as well as several related questions on relocating and settling into a new area.

For the majority, it seemed "quite precise to speak of their reactions as expressions of *grief*" (Fried 1966, 359). At their most extreme, these grief reactions were intense, deeply felt, and sometimes overwhelming. After moving, people reacted with statements such as: "I felt as though I had lost everything;" "something of me went with the West End;" "I felt like my heart was taken out of me;" "I had a nervous breakdown;" and "I felt like taking the gas pipe." Of those who had previously reported to have liked living in the West End "very much," 73 per cent showed evidence of extreme grief. Even 34 per cent of those who were ambivalent or negative about the West End grieved severely for their lost home area. The grief syndrome identified by Fried included vomiting, intestinal disorders, crying spells, nausea, and general sadness, depression. In Fried's cool words (1966, 359), expressions of grief "are manifest in the feelings of painful loss, the continued longing, the general depressive tone, frequent symptoms of psychological or social or somatic distress, the active work required in adapting to the altered situation, the sense of helplessness, the occasional expressions of both direct and displaced anger, and tendencies to idealize the lost place." Those symptoms spell bereavement, an issue that will be taken up in chapter 6.

While some relocatees may have welcomed the move, after two years, 46 per cent of the women interviewed still experienced a grief reaction.

This was often renewed when people were told that their former home had finally been demolished. Fried acknowledged the significance of relocation losses as they affect routines, relationships, and expectations. He found that grief associated with loss of place was closely linked to both loss of social network and of the physical structure and context of the dwelling. This grief proved strongest among those who had liked living in the West End, who were familiar with the greatest area within the neighbourhood, and who had lived there longest. In their grief, some families strengthened kinship ties, but many tried to remain close to the area they knew even though the personal relationships they had formerly enjoyed no longer existed.

Since the pioneering studies of Boston's West End, relocation pathologies have been reported from many countries under many types of political regimes. A study of four cities in the former Czechoslovakia found that 40 to 60 per cent of the residents were quite satisfied with what planners regarded as blighted housing (Musil 1972). When told that relocation was inevitable, only 17 per cent wanted their new dwelling to be located outside the neighbourhood. In Britain, extensive studies confirmed and extended Fried's work. In Hackney, for example, the community lost its sense of identity during redevelopment (Young et al. 1981, 61), and after Covent Garden was redesigned, people were also reported to have lost their sense of belonging (Anson 1981, 236). A series of studies of those resettled from inner cities to the sprawling low-income suburbs of Britain's larger cities confirm that the much-loved community feeling of the past could not be reconstituted in the new environment (Porteous 1977).

When residents are powerless, redevelopment projects ignore them or provide unsuitable mitigation. Clairmont and Magill (1987) show how the poverty of the Canadian blacks of Africville in Halifax, Nova Scotia, allowed them to be taken advantage of. Victims of the "liberal-welfare rhetoric of progress" of the 1960s, Africville's 400 residents were relocated between 1964 and 1967. The twenty-five-acre site is now occupied by Seaview Memorial Park, which overlooks Bedford Basin in central Halifax.

Africville people used kinship ties, race, community history, and their church as symbolic boundaries to distinguish their community from others around them. After being moved, they lost their community's historical ties to place, and were unable to accommodate extended families in the new residential environment. Feelings of powerlessness and apathy, perhaps akin to the fatalistic *destino* espoused by some West End residents, prevented effective resistance. In the long run, the city of Halifax felt that the relocation was not a success. For the evictees, it was a disaster. As Mrs. Hattie Carvery laments: "They took our homes ... the city moved us out of Africville in city garbage trucks ... they put us right down in the slums, a lot of us ... it was our settlement, our community. It was everything to us ... They offered us a wee little miserable bit of money for our homes, and then we were held with the threat over our heads that if we didn't oblige the city, they were going to bulldoze over us anyway." More profoundly: "The white people shamed Africville so much that ... a good many of the young people ... don't want to own their own heritage ... They made them ashamed. We weren't ashamed. We were proud" (Hartnett 1970).

The process continues. The rise of the "entrepreneurial city" (Harvey 1989) and the operation of the city as a "growth machine" demand constant redevelopment on an ever-larger scale. In societies without well-developed protection for citizens, urban redevelopment frequently involves violence, such as the mysterious fires that coincidentally burned down those sections of Singapore that the country's authoritarian government wished to redevelop for industry. In the 1990s, Singapore continued to dispossess its population from their homes (this time for hotels), leading, at last, to courageous protest movements. In Romania, Ceausescu's rural domicide policies, outlined in chapter 3, were matched by the wholesale demolition in the country's capital. Central Bucharest was virtually destroyed to create the Boulevard of the Victory of Socialism, leading to a People's Palace worthy of the Pharaohs. Even after the dictator's death in 1989, Bucharest's chief architect stated that he would complete the plans in an effort "to improve our image, and break with the mirage of the past" (Malcolmson 1994, 4). The psychiatrist Anthony Daniels asked Romanians about "the nature of an authority that could direct the efforts of an entire nation not only to the construction of something entirely without merit, but to the destruction of everything worthwhile from the past," and he answered his own question by divining that the demolition and rebuilding of central Bucharest was "intended as an advertisement, not of a product, but of men's individual insignificance in comparison with the power of the state" (Daniels 1991, 84–5). Similarly, the newly independent country (1991) of Turkmenistan has stripped the whole centre of the capital city Ashgabat of its housing, leaving only ministries, palaces, hotels, and building sites (Shihab 1997). And from totalitarian China, there were news reports in 1994 that told of boatloads of Chinese landing in Australia because urban redevelopment had destroyed their homes, only to be repatriated because "urban renewal and destruction of home is not a legitimate reason for getting refugee status" (Canadian Broadcasting Corporation 1994). It goes without saying, then, that urban squatters are given very little consideration when government and business decide to develop the land they are occupying (Hollsteiner 1977, Perlman 1982). In his novel *The Dreams of General Jerusalem*, Peter Marris (1989) dramatizes his experiences as a planner in an African city. Many of the significant characteristics of domicide are evident in this story of idealism thwarted by greed and ambition. As Marris notes elsewhere, slum clearance appears to be a form of social engineering, "not the replacement of dilapidated, insanitary, overcrowded houses, but the reformation of a way of life" deemed unsuitable by the elites (Marris 1974, 53).

Even in well-mannered Sweden, renewal projects are seen as a threat to a calm and safe life, and residents expected to move permanently experience a great sense of loss (Wikström 1994, 316–7). This being the case, it comes as no surprise that considerable resistance to urban redevelopment occurs across the globe, that opponents of "the politics of exclusion" refuse to consider housing as merely a commodity (Ransom 1996), and that the United Nations Habitat II conference, held in Istanbul in 1996, included a major debate between nations that see adequate housing as a laudable goal (India, the United States) and those that regard it as a human right (Japan, South Korea, and others).

Finally, it is no surprise either that urban redevelopment has been the subject of a considerable imaginative literature. The above-mentioned novel by Peter Marris (1989) shows how in Third World situations, planners, politicians, and aid agencies co-operate to violently oust poor squatters from their homes. André Brink fictionalized the evictions of District Six in Cape Town: "Dey turning Distric' Six into a smart place now, fo' de Gov'ment's White boys. We Coloureds getta kick inne arse en' it's out we go" (1974, 197). The story of Penelope Lively's *City of the Mind* is set against a background of London relocations "assisted" by extreme harassment of tenants by property owners who appear to lack certain emotions, "fiddly stuff like compunction, and vicarious distress, and compassion, and moral outrage" (1991, 64). The Hungarian novelist George Konrad "zeroes in on one of the uglier aspects of modern warfare, the undeclared war of city planners against city dwellers: the war of the manipulators of life against the livers of life" (Fuentes 1987). And when Samuel Becket was looking for strong expletives for *Waiting for Godot*, he resurrected, but finally rejected, the nineteenth-century Brussels curse: "architecte" (Bair 1978, 426).

Economic Development and Restructuring

Commercial and industrial changes, especially on city fringes or in non-urban areas, can result in significant loss of home. Again, land use changes brook least opposition in authoritarian countries such as China (Woon 1994), Turkey (Brazier 1996), and India. The displacement of people in India's Orissa state by coal mining, irrigation projects, and industrial expansion has been exposed in the tellingly titled *Depriving the Underprivileged for Development* (Pandey 1998).

Such economic change can bring domicide to whole regions. In Britain, many villages disappeared in the Tudor period as landlords converted tilled land to sheep pasture and extended their grand estates (Beresford 1955, Allison 1970). Four thousand enclosure acts reorganized six million rural acres between 1750 and 1880, increasing rents and forcing thousands of rural people into industrial cities (Chambers and Mingay 1966). The effects of such enclosures on the English poet John Clare is well known, for these drastic changes in his home landscape caused him mental agony and eventual insanity. Later still, the industrialization of family farmland made the Brangwens in D.H. Lawrence's *The Rainbow* feel like "strangers in their own place" (Pocock 1980, 345).

In Scotland, the Highland Clearances of the eighteenth and nineteenth centuries saw Scots chiefs also convert their land from arable cropland to sheep pasture, with little thought for the families made homeless. Indeed, the Clearances were in some cases particularly brutal, justified by the perpetrators on the grounds that they were, "upon the whole, advantageous to the nation at large" (Prebble 1969, 106). Whole islands in the Hebrides were cleared of people, the houses burned, and the former inhabitants clubbed, fettered, and thrust onto emigrant ships. Similar regional depopulation took place in Ireland (Orme 1970), where the English engaged in large-scale geopiracy. Deportees were said to live "more pleasantly abroad than at home" (Strauss 1954, 149), the usual justification for slum clearances and evictions. Irish evictions figure prominently in novels (Uris 1976). In remote areas of the Celtic Fringe, landlord power remains strong. In the early 1990s, a Swiss creditor bank foreclosed on the absentee owner of the Scottish island of Gigha, but prospective buyers were not informed of the 140 people for whom they would have semi-feudal responsibility (Hancox 1992, 23).

Similar "clearances" are going on today as economic demands lure corporate timber, oil, uranium, gold, and other interests into the world's once-remote deserts, tundras, and forests. Such developments are especially detrimental to indigenous groups, which are strongly attached not to a type of terrain, but to a particular place or set of places. Trimble (1978) has studied a number of such cases, particularly in Latin America, and concludes that relocation is invariably initiated by an agent, industrial and/or governmental, which is external to the group, and it is implemented for economic, political, and other reasons. The programs are conducted in such a way as to give the appearance that the external agent holds the cultural preservation of the relocated group as a primary objective. However, the motive for the move is actually the exploitation of the lands that the relocatees are forced to leave, and little or no effort is exerted by the external agents to prevent the deleterious psychophysiological and psychosocial effects generated by the move. In a later study of four cases, Trimble (1980) adds that they were all native, indigenous groups with cultural orientations and lifeways quite different from those who forced them to resettle. They were relocated so that their lands could be exploited for the common good, to benefit a much larger population. None of the four groups volunteered to resettle, and after the resettlement, none of the groups benefited directly or indirectly from the development of their ancestral lands. Each of the groups experienced hardship as a result of domicide and resettlement.

Such conclusions are an overwhelming indictment of the way in which metropolitan populations exploit and devastate indigenous peoples. For, in some cases, as with the Kreen-Akrore of Brazil, domicide means ethnocide, the death of a people and its unique culture. As Barnabas and Bartolomé (1973, 7) note of the resettled Mazatecs and Chinantecs of southern Mexico, their culture vanished after removal. "For others, the transfer meant death. At least 200 simply died of depression (*tristeza*); the removal was especially hard on the aged, who grieved upon leaving lands where ancestors were buried and their sacred objects secure."

Less traumatic, but still a potent source of domicide, is the creation and closure of mining towns. Abdelrahman Munif's novel *Cities of Salt* (1987, 106) speaks of the changes wrought in Saudi Arabia by twentieth-century oil exploitation. One character recalls "The long-ago days, when a place called Wadi al-Uyoun used to exist ... and a brook, and trees, and a community of people." The former inhabitants felt that their village was now "a cruel, wicked sight that resembled death." They were relocated, only to ask: "Can a man adapt to new things and new places without losing a part of himself?" (Munif 1987, 134). The experience of northeast England, where numerous coal-mining towns lost their *raison d'être* because of changes in government policy after the advent of Margaret Thatcher, demonstrates how rapidly whole regions can be wrecked (Hudson 1989).

Most vulnerable of all are mining communities established in remote locations in the hot and cold deserts of resource frontiers such as northern Canada, Chile, and Australia. These "single-industry" or "company towns" are indeed "communities on the edge" (Randall and

Ironside 1996), both geographically and economically, for once their resource is no longer viable to mine, the *raison d'être* of the town disappears, and closure must be considered. As there are approximately 700 resource-dependent towns in Canada, the problem has been addressed most frequently in that country (Bowles 1981, 1982; Brookshire and D'Arge 1980).

Bradbury and St. Martin (1983) investigated the "winding down" process in the town of Schefferville, Quebec, in which the iron mine was closed down in 1982. Their analysis points to the parallel processes of decreasing corporate involvement (withdrawal from public service provision and municipal affairs, and disinvestment) and community winding down (emigration, instability, rumours, social dislocation). The authors agree with Bluestone and Harrison (1980) that large modern corporations, and especially conglomerates, feel no compunction in closing down even profitable operations for a variety of reasons relating to centralized management and control. They found the prevailing atmosphere among the inhabitants of declining or dying one-industry communities to be one of rumour, uncertainty, and anxiety. This is often due to failure to communicate on the part of the company, an insensitivity toward labour which merely exacerbates "residents' feelings of neglect, as well as their feelings of impotence in the face of changing circumstances resulting from changes in the company's structure [and] the absence of local participation in decision-making" (Bradbury and Sandbuehler 1988, Bray and Thompson 1992). In *Coping with Closure* (Neil et al., 1992), comparisons are made between Canada, Scandinavia, and Australia. Given that the life of an ore body can be predicted, there is no reason that the complex process of town closure cannot also be anticipated, with advance notice given, relocation benefits anticipated, and counselling services provided. In some cases, mine-site towns need not be built at all, for local indigenous labour could be employed, or fly in – fly out labour-commuting programs established. Alternatively, state policies could work toward retaining redundant towns via the long-term establishment of alternative sources of employment. With effective long-range planning, there is no reason for the tragedy and trauma that so often accompanies the closure of isolated company towns.

Remote Pacific islands have suffered much more tragically than mining towns in industrialized countries. The cases of Nauru, Banaba, and Bougainville, destroyed by economic exploitation, may well be compared with the military annihilation of other Pacific islands recounted in chapter 3.

The Republic of Nauru in Micronesia, is one the smallest nations in the world with about 10,000 people living on a single island with an area of only 21 square km. In 1899, it was found to contain huge deposits of guano phosphate, which were mined by a British, Australian, and New Zealand consortium until the near-exhaustion of the resource in the 1990s. The interior of the island became a scarred landscape devoid of soil, a wasteland of steep coral pinnacles. Add to that an almost exhausted phosphate resource, an ugly landscape unlikely to attract tourists, the possibility of a barely profitable coconut products industry, and the need to import most food and all fuel, machinery and manufactured goods, and you have a rather depressing scenario.

Yet, Nauruans rejected resettlement elsewhere. They were concerned that relocating to Australia would cost them both their identity and their sovereignty over their abandoned island. Moreover, they did not wish to relocate to the already-inhabited island they had been offered, fearing resistance and possible racial disharmony. Finally, they were well aware of the resettlement problems of the Banaban people in Fiji (which we discuss further below). By the 1990s: "Thanks to human avarice, greed, and shortsightedness, our island is mostly a wasteland. Phosphate mining ... has reduced the island, except for a coastal strip, to a desert of jagged coral pinnacles, uninhabitable, unusable ... Our land was literally exported ... to fertilize the fields of the industrialized world" (Clodumar 1994, 3). Despite this unenviable situation, resettlement was rejected, leaving only the option of landscape rehabilitation.

In 1993, Nauru successfully claimed compensation from the former exploiting countries for the indiscriminate destruction of their island's land surface over a period of sixty years. Using royalties and the compensation package, Nauru is proceeding with a twenty- to forty-year rehabilitation of the island in three overlapping and interlocking steps: physical, biological, and cultural. Asked if such a tremendous operation was worthwhile, a government spokesman said: "Islanders know, better than anyone, land is both the basis of human life and also limited. Nauru is our home, our only home ... Nauruans do not emigrate. Our people are devoted to the land even in its present sorry state" (Clodumar 1994, 5).

Although their whole land base is ruined, Nauruans cling to their island home. Nearby Banaba, also known as Ocean Island, suffered a much worse fate. Against the Banabans' wishes, the British Phosphate Commission mined the island's guano rock from 1902 to 1979 in "perhaps the best example of a corporate/colonial rip-off in the history of the Pacific Islands" (Stanley 1993, 25). In 1916, the island's status was changed from a protectorate to a British colony, so that islanders could not withhold their land from mining. In 1928, Sir Arthur Grimble, famous for his sentimental South Sea yarns, expropriated the whole land surface. In 1942, the Phosphate Commission purchased the island of Rabi in Fiji, 2,600 km away, planning to resettle all 2,000 Banabans there. They were relocated to Fiji after 1945, and today, they have lost most of their Micronesian culture. Only after twenty bitter years of court battles were the Banabans able to obtain compensation from the reluctant British government; some returned to their island, but their claim to Banaba remains uncertain because the island is now part of the Republic of Kiribati. With the usual disregard of industrial nations for small islands, an Australian project in 1991 proposed to use the mined-out island as a dump for liquid chemical wastes.

In the case of Bougainville Island, off the east coast of Papua New Guinea (PNG), a major copper mine was opened in the early 1970s. To accommodate the operation, several new towns were built and "old villages" were relocated (Cummings 1972, 55). The mining company was yet another subsidiary of the giant Rio Tinto Zinc conglomerate. It provided 3,000 jobs, created an open pit six miles wide, and flushed its tailings down the Jaba river to the sea. Royalties appear to have benefited the PNG mainland more than Bougainville itself. By 1988, active resistance to the mine and the PNG government had emerged, fuelled by separatism, environmentalism, and liberation theology. In 1989, the Bougainville Revolutionary Army took over the island, expelled the company, closed the mine, and attempted to destroy all foreign physical structures, including schools, factories, power stations, houses, and plantations. Unlike Cambodia's Year Zero, this return to a past way of life was probably feasible, for the traditional subsistence cultures had been disturbed for only two decades. PNG government troops at once blockaded and then invaded the island causing considerable loss of life. In 1997, the "security situation" was still deemed "inappropriate" for extensive rehabilitation of the island (Australian Agency for International Development 1997, 25), but some rehabilitation had begun (Francis 1998).

The Melanesians of Indonesian Papua New Guinea (Irian Jaya) have been equally badly treated by central governments and mining companies. In 1967, the Indonesian government authorised the New Orleans-based Freeport Company to open a copper and gold mine at Ertsberg. Local highland people were not consulted, but their huts were razed and a village built for them on the hot malaria-infested coast. Keen to assimilate Papua into the Indonesian economy, the Indonesian government also obtained World Bank support for their Transmigration Program, whereby people from densely-populated central Indonesia were transferred to the "vacant lands" of Papua. Between 1984 and 1989, about three million hectares of forest were opened up to Javanese settlers, forcing traditional Papuans from their homelands and thus creating a pool of cheap, landless labour.

Papuan resistance began in the 1970s with the guerrilla operations of the Oposisi Papua Merdeka (Free Papua Movement). Such was the Papuan opposition that when the Ok Tedi mine was begun in 1982, the company was compelled to recognize indigenous land rights and the Indonesian government forced to renegotiate local royalties from 5 to 20 per cent. The severe land degradation caused by Ok Tedi, however, led Papuans to launch a lawsuit against Broken Hill Proprietary Ltd. in its home town of Melbourne, Australia. Similarly, a suit against Freeport was brought in New Orleans in 1996, alleging environmental and cultural destruction, while the Bougainville islanders filed a claim against Rio Tinto in California in 2000 (Pallister 2000). Thus, the transnational reach of mining companies is now being matched not only by environmental organizations, but also by villagers themselves. In Papua, landowners have come to wield increasing power, extracting compensation from both governments and companies, bringing some operations to a standstill, and preventing others from beginning. Nevertheless, even successful resistance radically changes traditional ways of life.

While some Melanesians are clearly willing to take up arms to prevent domicide, this response is less appropriate in the Western world. Nevertheless, in industrialized countries, there are many cases in which peoples' homes are destroyed for industrial and commercial redevelopment, and working-class people may be treated almost as badly as the Micronesians of Nauru and Banaba.

In the book *Planned to Death* (1989) I (Porteous) describe the slow murder of Howdendyke, East Yorkshire, the village in which I grew up. Its demise was begun by Thatcherite restructuring, the creation of unnecessary port installations to "discipline" the unionized port of Goole nearby. Since 1969, port development has grown in several phases, each one of them taking more land once occupied by houses. Abandoned by planners and politicians as a "dying village," Howdendyke is not dying, but rather being slowly killed off by piecemeal development. In 2001, after about thirty years of slow replacement, the village is only one-third its original size, has lost much public access to the river, is heavily polluted, and has seen the demolition of its pub, shop, and village hall.

All this came about without consultation with, or opportunity to comment by, the inhabitants. Business corporations, district politicians, planners, and other officials co-operated in condemning village houses, demolishing them, and removing their residents to a nearby town. Five conditions operated to produce this stealthy death of a thousand cuts.

Land, housing, and employment were controlled by distant corporations with no interest in the village other than as an industrial site. Previously lulled into complacency by a paternalist regime, most of the villagers were tenants dependant on the companies for both housing and employment. Furthermore, they were predominantly working class, with no knowledge of planning and no experience of effective resistance. Thus, it was not difficult for the district planning department to ignore county planning guidelines and public participation requirements. Finally, district politicians opted to sacrifice Howdendyke in favour of "wider interests," such as increased regional employment.

Collective action came too late to save the village; by 1998, the riverfront was "swept clean" of houses, ready for future port development. The range of emotions found by Fried (1966, 1982) in Boston was much in evidence, however, both among those who were relocated and those who remained in the village. Anger, resentment, sadness, grief, and a sense of loss were widely felt. Blame was also assigned, albeit in oblique ways (Porteous 1989, 188): "It seems that certain people in high places are doing their utmost to drive us out;" "The village was here before the factories. We wish to stay here, not to be hounded out by the industries around us;" and "We feel we are being squashed out of existence from the place we have lived in all our lives." To which the companies replied that "It is in the nature of businesses to grow," while both politicians and planners reminded villagers that the destruction of their homes had been perfectly legal.

In such a sad scenario, one may be heartened by individual examples of resistance. Living alone in one of only two occupied houses in a partly derelict row of nine, widowed Mrs. Vera Arnold resisted all suggestions that she move to a nice, clean, modern flat in a nearby town. "This is my home," she said. "We were happy here. All my memories are here. I'd never want to move." And, " You see these houses flattened and then for years brick rubble and the land left desolate and nothing done. It's a wicked shame, especially when people want to stay." Resisting to the last, Vera remained in her home until her death, aged seventy-eight, in March 1998.

MAJOR PUBLIC FACILITIES

> My name is Ozymandias, king of kings:
> Look on my works, ye Mighty, and despair!
> Percy Bysshe Shelley, "Ozymandias" (1941, 251)

There is a considerable literature in geography, planning, and engineering on the location of public facilities and infrastructure. A good deal of this is summarized in Massam's *The Right Place* (1993), which seeks to integrate two rather distinct approaches: highly theoretical location theory involving location-allocation modelling, and the rather more humane preference-based decision analysis. The aim of the latter is to invoke "shared responsibility" as an ideal in the creation of "socially acceptable collective choice planning procedures, which could be used to tackle controversial and complex public facility location problems" (Massam 1993, xv). Ultimately, however, the final choices are likely to be made by experts informed by highly complex computer-assisted decision-making models and approved by politicians seeking their own ends. Unfortunately, input into the decision models, however technically perfect, is seriously compromised by the underlying differences in values that separate conflicting interest groups (Obermeyer and Pinto 1994, Malczewski 1995).

Attempts to deal with this problem have included a Competing Values framework, which looks at critical perspectives as being consensual (decisions must flow from full participation by all parties), rational (solutions must be efficient), empirical (decisions are based on data and thus accountable), and political (decisions must be legitimate). Not surprisingly, the weight attached to such values varies greatly according to the cultural context (Vari et al. 1994). In France, for example, public participation is weak compared with its vigorous counterpart in Canada (Massam 1995).

Our aim here is not to engage in a theoretical debate on public facility siting, but to muddy the waters still further by emphasizing the claims to be heard of those who are so often disregarded – the people whose homes and lands are taken from them in the public interest. Even the most sophisticated technical and judicial inquiry systems may falter before the homely voices of ordinary individuals who "chose our home for a lifetime" or "laid every brick ourselves."

Domicide in public facility location is discussed in terms of roads, airports, national parks, and, most important of all, dams and reservoirs.

Roads

From the 1950s to the 1970s, the richer countries of the industrialized world carried out an enormous program of road building. It was a boom period for land developers, oil companies, vehicle manufacturers, concrete and asphalt makers, and their political friends. It was the OPEC oil crisis of 1973, rather than common sense, that finally slowed down the submergence of non-urban land beneath the concrete glacier.

Transportation planners of the 1950s generated their plans almost wholly in terms of material costs, engineering feasibility, and political desirability. Unfortunately, at that time, the political process in planning allowed next to no involvement on the part of citizens, who sometimes would wake up one morning to find their homes expropriated for a new inner-city expressway. Too often, new city highways were planned to cut great swathes through working-class and ethnic districts, though their benefits went mainly to outer suburban commuters. In the 1960s, planners, politicians, and the public had to learn the hard way the unwelcome truth that: "the planning and construction of transportation facilities are far more than an engineer's technical problems. [They present] important political decisions that affect, directly and indirectly, the lives of millions of people, both users and persons impacted upon by the construction" (Lupo et al. 1971, 171). In other words, new highways generate not only new traffic that fills them beyond capacity, but also domicide.

One of the problems for residents "in the way" of major highway projects is the inordinate time it takes to move from approval of the project, through compulsory purchase, to the demolition of dwellings, to the actual construction of the new road. A report by the British National Audit Office (Comptroller and Auditor General, U.K. 1994) revealed that 60 per cent of dwellings purchased by the Department of Transport for road schemes outside London were not needed, and that one-quarter of all the houses bought up are left empty for years, thus causing planning blight in the neighbourhood. The report had little to say about the feelings of those whose homes were expropriated.

The renewed expansion of British road building in the 1980s, in part to accommodate enormous European trucks, led to mass protests on the part of those affected. The NIMBY (not in my backyard) syndrome makes good sense at the level of the individual homeowner or small village: in 1983, the M42 plowed east-west across the "backyard" of the village of Curdworth, Warwickshire; ten years later, a planned "relief road," running north-south through the village's "front garden" was announced. Even elderly conservative ladies felt the need to lie down in front of the bulldozers. Said one of them: "We'll be in the jaws of a nutcracker. We're not going to let this happen again" (Brooker 1993, 11).

Those wishing to save their homes have formed alliances with environmentalist groups with slogans not NIMBY but BANANA (build absolutely nothing anywhere near anyone). Other groups, such as the Council for the Protection of Rural England, ask how we may weigh the convenience of the automobile against the destruction of irreplaceable historic landscapes that are part and parcel of the national identity. Still others campaign against the destruction of habitat – the homes – of rare birds and mammals. Throughout the 1990s, stop-the-road battlegrounds erupted across Britain (Road Alert! 1998), but as early as 1975 a popular comic novel based on a motorway construction scheme had appeared – *Blott on the Landscape*. Others might well agree with one of its character's assessments: "... when you have been in the public service as long as I have, you will know that Inquiries, Royal Commissions, and Boards of Arbitration are only set up to make recommendations that concur with decisions already taken by the experts" (Sharpe 1975, 226). The image of Arthur Dent protecting his home by lying in front of a bulldozer at the beginning of *The Hitchhiker's Guide to the Galaxy* (Adams 1979) has become a public icon.

Loss of home to highway construction is therefore a common experience. In New York, Berman, victim of an expressway, eloquently mourns the loss of his home in the Bronx and seeks "to generate a dialogue with my own past, my own lost home, my own ghosts" (1982, 342). Such experience in the United States in the 1970s led to a considerable understanding of the effect of relocation caused by highway construction. Finsterbusch summarizes this and concludes that although most people will adjust well, there are always some who suffer psychological stress in attempting to come to terms with all the "severely traumatic" changes, and some who will die (1980, pp. 112 ff.).

The examples outlined here confirm Finsterbusch's findings since they demonstrate that some people may adjust well and achieve a better solution (higher cash settlement, stopping the project or creating a better home for themselves) through effective protest, but for others, the effect of domicide on their lives and their families will be significant. Recently I (Smith) met Alice Hambleton, a musician. When I told her of my work, she wrote me the following letter that illustrates the means, motives, and effect of domicide from the perspective of someone who lost her home not to a highway, but to a parking lot at a ferry terminal at the end of the longest highway in the world, the Trans-Canada Highway (Hambleton 1994). As will be discussed in later chapters, the expropriation process and achieving a fair price for property rights has not always been supported by adequate legislation, policy, or process, and the system certainly left much to be desired in Hambleton's case. While there is no doubt that compensation value was a serious matter for Hambleton's parents, of greater importance here is simply to listen to Hambleton speaking about the impact on her family, and how they perceived the common good rationale that justified the construction of the parking lot, the lack of consultation, and the behaviour of government agents.

In 1963, I was five years old. My family was living on Marine Drive, just above Horseshoe Bay, in West Vancouver. My parents had a mortgage on a large, three-storey duplex (which my dad had built), right down in Horseshoe Bay. My parents had both sides of the duplex rented out to tenants.

Horseshoe Bay was a small picturesque village. The Bay was a fun place to live in those days.

One evening, one of the renters from the duplex came to see my parents on Marine Drive. He was angry because a government agent had knocked on his door and told him that he would have to move soon. The agent said that the government would be buying the duplex for the road. My parents were surprised. This was the first they'd heard about it.

About two weeks later, the government agent came to my parents' house on Marine Drive and offered them a price for the purchase of their duplex. My parents refused the offer. They didn't want to sell the duplex. And so the government agent stormed out. My parents were left in a quandary. The tenants moved out of the duplex and the duplex sat there empty with the next mortgage payment due. So, my parents decided to move us all into the duplex and sold the house on Marine Drive.

So, for the next four-and-a-half years, my family lived in the duplex in Horseshoe Bay. Meanwhile, all the houses around the duplex were sold and destroyed. We were now the last house on the longest highway in the world. And for four-and-a-half years, our house was the only thing that was in the way of W.A.C. Bennett's big Expanded Horseshoe Bay Ferry Terminal dream.

The government agent never came back. But the government hired two watchmen with guard dogs to patrol the properties that they'd bought around us. They also did their best to intimidate and torment my family, to try to make us move. Predictably, the watchman would only torment us kids and my mom. He was too "chicken" to do something like that when my dad was around. So this is what the government did to those citizens who wouldn't cave in to its unreasonable demands.

The government watchman constantly patrolled around our house with his dog. He'd stand in our driveway for long periods of time and just watch us, menacingly. At night, he'd park across the street in his car and watch us through our windows. This went on for more than four years, until my dad finally called the police and said that he was going to shoot the watchman if he didn't stop bothering our family.

A short time later ... about two weeks before Christmas 1967, my parents received their eviction notice. We had to be out by the end of December, just a few weeks away.

My parents went to a lawyer. He said that they couldn't fight the government. As a result, our family was expropriated, in the middle of winter, at Christmas time. My parents only got 2,000 dollars more for their property than they had originally been offered. The government was supposed to only evict property owners for a road, but our house and yard became the Horseshoe Bay pay parking lot, instead. My parents are *still* upset about this ... twenty-five years later.

My mom thinks that if the government is going to expropriate property for a road, they should at least offer fair compensation. They should settle it through arbitration, until both parties are satisfied. Also, they should give proper notice and should complete the process within a reasonable time period. The government has no right to torment or financially handicap families who don't want to sell their property. And something like this should never drag on for five years. After all, when the government finally needed the property, they just took it.

The expropriation of our house resulted in almost five years of tension between my parents and the government. The tension affected our lives, and it affected us kids. We all grew up mistrusting the government and feeling a general contempt for "the system" and for "authority." We all became "rebels." When I go to Horseshoe Bay now, I feel nostalgia for what was, and is now, just a memory. I wish I could see my house and yard again and see Horseshoe Bay the way it used to look. I also feel sad because the most beautiful place in the world got turned into a major tourist terminus, where the biggest business in town is parking. My whole family feels cheated. Our life in and around Horseshoe Bay was taken away without our consent.

Unlike the previous example, one North American protest against highway development that did make a difference involved the community of Durham, North Carolina, in which a black neighbourhood known as Crest Street was threatened by an expressway proposal in the 1970s (Rohe and Mouw 1991, 60 ff.). The project was pushed by a politically well-connected city councillor and the local business community, including both black and white business leaders, since it would open up a route to the suburbs and to a research and manufacturing park. During the first phase of the project, several neighbourhoods were to be razed, the first being a black, low-income area called Hayti. The second area to be demolished was Crest Street, which, due to its imminent destruction, had been allowed to deteriorate.

The black community of Crest Street saw that people uprooted from their homes in Hayti had been placed in housing conditions far worse than their original homes. The community also found ready allies in young white community activists who opposed the expressway because it would come close to their homes. In studying the project area, it was found that the average period of residency in the area was just over thirty-six years, that 65 per cent of the community's residents had relatives in the immediate area, and that 90 per cent of the residents believed the community to be a safe place to live. It was this information that proved crucial to the neighbourhood's opposition to the expressway. In the end, although the community was relocated to an adjoining area that was much improved, it moved on its own terms, and, where possible, existing homes were moved to new locations. The domicide that occurred to the community of Hayti was prevented in Crest Street. Resistance proved effective. At times, it was also effective when confronting proposed airport developments.

Airports

Airports, almost always located on the peripheries of cities, take up large amounts of strategically located land. Whereas urban road projects can unhouse thousands of citizens, new airports usually discommode only small numbers, but they nevertheless are sometimes capable of fierce resistance. Two of the more interesting examples are those of Narita Airport, near Tokyo, and the projected Third London Airport, which as yet has failed to come to fruition.

Japan is known for a decision-making process involving consensus rather than confrontation, but that cultural trait appears to have been ignored by the authorities when plans for Narita airport, forty miles from Tokyo, were announced in 1966. Assuming that Japanese farmers would be proud and patriotic enough to sell their land without legal expropriation measures, the government simply unveiled its blueprint without consulting the 360 farming families affected.

Some farmers, however, were deeply attached to their land and resisted government blandishments. Radical environmentalists and other protesters came to their aid, and legal battles were matched by physical battles with the authorities in which several people were killed. Although the government forcibly expropriated the land of 350 households, who were uprooted after 1970, a handful of farmers resisted. Holding on to strategic pieces of land and publicly challenging the authorities to destroy their crops, the farmers found public sympathy through media exposure by appealing to positive Japanese cultural attitudes about farmland. Although one runway was built, they were successful in preventing the completion of a second. Thus, the world's sixth-busiest airport languishes with only a single runway, although government plans to build two more are frequently revived. After over thirty years of resistance, only four families were still holding out in 1997. They have resisted force, persuasion, temptation, and harassment to keep their land. Koji Kitahara's dream remains: "stopping the airplanes, tearing down the buildings, and removing the runways" (Wudunn 1997, A11).

Encouraged by this long-running example, airport plans have been questioned across the industrial world. The third Paris airport at Roissy and the second airports in Montreal (Mirabel) and Toronto (Pickering) in Canada also led to resistance. Perhaps most spectacular in terms of tunnels and treehouses was the late 1990s direct action against Manchester Airport's second runway project. With the British public both sensitized and sympathetic because of the fairly successful anti-road campaigns of the previous decade, as well as the legitimization of protest action on soap operas such as *Coronation Street*, considerable support was given to the protesters, once again a motley coalition of "small-town fuckwit Tories, the green welly brigade, *Guardian* reader advertising types, village preservation sorts, and the 'this is going to knock 30 per cent off the value of my house' lot" (Aitkenhead 1996, 5). Despite a package of environmental mitigation promises, opponents stressed the point that what will be lost is a landscape of human identity, "the particular, the individual, the unique" (Evans 1997, 24).

From the point of view of potential domicide successfully resisted, the project for London's third airport repays our attention. The proposal for a third London airport was subject to the most elaborate of inquiries, which lasted for over two years between 1969 and 1971, and which cost £1.12 million. It is given detailed attention here because the author (Smith) was involved in the inquiry and therefore has a greater sense of the destruction the contemplated project would have meant. The proposal was a product of its time, a period when an apparent exponential growth in air traffic could easily be coupled with British pride in the importance of having a world-class international airport. To quote Justice Roskill, who led the inquiry: "The hostile jibe during the second world war that this country was no more than an aircraft carrier should in the last thirty years of the present century be a source not only of pride but of economic and political strength" (Commission on the Third London Airport 1971, 5).

To give the Roskill Commission its due, the siting of a third London airport was an extremely difficult problem, and the inquiry was thrown into the middle of a paradox of progress. Air travellers, and those who provide air service, want a facility as close to major population centres as possible. To do this normally means the destruction of homes or, at the very least, significantly increased noise pollution. The inquiry received evidence that by the year 2000, one hundred million passengers would use a new airport. No one argued against the need for a third London airport.

The commission decided to make use of cost-benefit analysis to aid systematic decision making. It recognized that cost-benefit analysis could be criticized for avoiding the real issues around the sacrifice of homes and peace and quiet but felt that this method ensured "that decisions are taken on the basis of people's individual values and choices as revealed by their behaviour rather than on the decision-makers' own preferences or standards or those of vociferous and politically powerful groups" (Commission on the Third London Airport 1971, 12). Such reliance on rationality was seen by one critic as "the culmination of one of the dominant trends of the political sixties; the conviction that the rational and the efficient, rather than the picturesque and the sentimental, must prevail" (McKie 1973, 15).

In studying the reaction of people who would have been affected by the final choice of the commission, the resistance displayed by the inhabitants of Cublington is exemplary. The choice of Cublington was, in the eyes of local historian David Perman (1973, 13), "an attempted rape of a section of English rural life that was repulsed with the stubborn and inventive resistance that the English have traditionally shown to foreign invaders." The choice was made primarily in the belief that Cublington made the best economic sense as an airport site. Had the project come to fruition, it would have destroyed three villages – Cublington, Stewkley, and Dunton – and possibly two others – Soulbury and Whitchurch. One thousand seven hundred people would have lost their homes, and 10,000 people would have been affected in other ways.

At the local hearings, 811 people sought leave to speak; there were 203 actual witnesses and 725 letters of protest. Excerpts from one letter are a record of the hopes and fears of the people of the area: "I am 82 years of age, and I was born in North Bucks and I shall not leave here until I am compelled to do. We chose our home for a lifetime. We like it. We like the generous rooms and layout. We like the small enclosed garden which is green and mature, with flowering shrubs and lilac hedges ... We are not affluent people. We paid £5,700 for our house, and the rest has been achieved by sweat ... this is my home, and I won't leave here except by force" (Perman 1973, 100). Following a highly effective local resistance, Cublington won its reprieve on April 26, 1971, mainly because of the growing concern about conservation of the environment, although the protest of those who might lose their homes must not be forgotten. The dominant social class represented in the primary protest group (Wing Airport Resistance Group) was lower middle-class, people who simply could not afford to lose their homes (Perman 171). Further, it is also clear that much of the environment lobby, so roundly criticized by Pepper (1980) as "ecofascist," was actually protesting against the loss of landscapes that profoundly express the English identity. They are "an organic part of the permanent heritage of England" (Pepper 1980, 178).

During all this turmoil it was discovered that air traffic was not increasing as rapidly as had been forecast. As Barfoot (1971, 33) remarks: "What did the Roskill Commission produce, other than frustrating delay, a wrong decision, and heartache for many in Buckinghamshire?" With the third London airport idea scuttled, the British government announced that Heathrow and Gatwick were favoured sites for a new runway within the next twenty years (Harlow 1993, 9). If Heathrow were chosen, over 3,000 houses would have to be demolished as well as fifty-five community and recreation buildings, eleven public buildings, and ten hotels. If Gatwick were chosen, one hundred homes around the airport would be bulldozed, and the Tudor village of Charlwood would be sandwiched between two runways (Elliott 1993). In planner Peter Hall's words: "The story of the third London airport is an extraordinary history of policy reversals, last-minute abandonments, contradictions, and inconsistencies in forecasts. It has all the ingredients of a great planning disaster"(Hall 1980,15). The story has now reached an ending of sorts as Stansted in Essex, by quiet growth, has become the third London airport.

National Parks

Unlike roads and airports, the advantages and disadvantages of which are so obvious, the creation of national parks at first seems to be a wholly benign policy. Such parks are developed to conserve landscapes and ecologies, for aesthetic reasons (Porteous 1996), and for public educational and recreational benefits. We are not likely to be aware that at least one popular model of park planning can result in extensive domicide.

Although the prevailing model for developing national parks is nineteenth century and American (Tripp 1998), the English had long experience of creating private parks before their national park model was established in the twentieth century. Medieval royal hunting parks were cleared of people to make way for deer, and in Tudor times, the "emparking" of land became so widespread that the peasantry found their lands, hamlets, and burial grounds disappearing behind the park pale, and themselves left to wander the roads as homeless and landless day labourers (Lasdun 1991). Henry VIII set a fine example by razing the whole village of Cuddington in Surrey to create a park for his palace of Nonsuch. Between the years 1670 and 1720, more country houses were built in some parts of England than in any other half century, and each needed a park. Thereafter, park making became an aristocratic English pastime until the early nineteenth century, and countless villages were destroyed and roads diverted.

Although the literate generally applauded these "improvements," we hear little from the nameless victims they spawned. In contemporary literature, however, we find considerable sympathy with the plight of the dispossessed. Oliver Goldsmith's *Deserted Village*, though sentimental, lays the blame for domicide squarely where it belongs: "The man of wealth and pride/ Takes up a space that many poor supplied" (1996, 11, original 1770). And Wordsworth's early poetry, particularly *A Night on Salisbury Plain*, is filled with images of home found and lost. The nameless Traveller and the wandering Female of this poem talk about "The loss and apparent irrecoverability of a permanent local habitation" (Janowitz 1990, 96) in a world of political, economic, and social oppression. When the person from Porlock broke into Coleridge's reverie, he might have interrupted to some purpose had he asked how many homes on the banks of the river Alph were destroyed to create Kubla Khan's Xanadu.

Nevertheless, it was this history of reverence for the English countryside that was translated, in the twentieth century, into the English model of the national park. From being a landscape of power, the English countryside became a landscape of the empowered as town dwellers fought for access and finally succeeded, after the Second World War, in establishing a series of national parks in northern and western England and Wales (Short 1991). Above all, it was an ideology of national parks as living, working landscapes that triumphed in Britain. Unlike the experience of earlier private parks, people were not removed from twentieth-century national parks. Indeed, farmers in particular were encouraged to remain and subsidized to continue traditional ways of life, which, in turn, preserved cultural landscapes considered essential to both regional and national identities.

Ideology proved very different in North America. Parallel with the belief that the American wilderness had to be tamed and converted into a fruitful garden was the paradoxical ideology of wilderness as Eden – an Eden that must be preserved without change for its aesthetic, scientific, religious, and national values (Short 1991, Porteous 1996). To love, protect and even worship nature was a deistic goal worthy of Wordsworth, Emerson, and Thoreau (Graber 1976). Religion and nationalism combined in the *Yosemite Act*, passed by Congress in 1890, which created Yosemite National Park, forerunner of the American national parks system.

This late nineteenth-century romantic American view of the national park emphasized "pristine beauty and wildness," and although us artist George Catlin in 1833 suggested that future national parks might contain "man and beast" (Mitchell 1981), the us National Parks Service soon reinterpreted this dictum in their clear preference for beasts over humans. Indeed, the prevailing twentieth-century model of the national park soon came to mean a wild area as free as possible of residents. While this notion proved quite practicable in much of western North America, it was much more difficult to implement elsewhere. Yet, internationally, the American wilderness model of the national park has triumphed over the English cultural landscape model.

Clearly, any attempt to establish a human-free wilderness national park in a populated area will inevitably lead to the need for population removals. Like most Western concepts applied without cultural sensitivity in the Third World, the wilderness park ideal has had serious implications for the subsistence needs of existing populations of cultivators, herders, and hunter-gatherers who suddenly find themselves within national park boundaries. Rao and Geisler (1988, 210) provide a number of examples including the San, a group of hunter-gatherers in Botswana who were resettled for the creation of the Gemsbok National Park; the Masai, pastoralists who were removed from their lands by the creation of the Nairobi National Park; and the Ik, a nomadic tribe from the Kidepo Valley of Uganda. The latter case is horrific.

The Ik had been hunters and gatherers, living a nomadic existence throughout the borderlands of Uganda, Sudan, and Kenya. Just before the Second World War, they were encouraged to settle in northern Uganda. These boundaries were hardened following designation of Kidepo National Park. Prevented from hunting in the park and punished when found "poaching," the Ik were compelled to learn to farm, for which their mountainous country proved quite useless. The 2,000 Ik called themselves "the mountain people," and their attachment to their mountain homeland was so strong that they resisted all administrative attempts to relocate them. Obeying the law but passively wondering at the power that ordains that animals shall be preserved while humans may perish, they preferred to die of starvation rather than move.

The Western world was shocked when Turnbull's *The Mountain People* (1972) described what happened to Ik society under these desperate circumstances. Severe lack of food caused the collapse of traditional social relations and the utter disappearance of family love, friendship, or neighbourly compassion and assistance. Even the essential nurturing relationship between mother and child attenuated; all interests were focused on the individual stomach. "There is simply no community of interest, familial or economic, social or spiritual," wrote Turnbull (1972, 157, 295) "The Ik are ... a people without life, without passion, without humanity."

Interestingly, well-known reviewers of the book, including Margaret Mead, Ashley Montague, and Sir Julian Huxley, focused upon the degradation of the Ik, an apparently monstrous people living fearful, selfish lives, and dying out because of the abandonment of their humanity. Much was made of the Ik as a possible forerunner of collapse and social decay in our own selfish industrial society. Little was said about the "needs" of Western tourists for game parks, and of the Ugandan government for the consequent revenue, which had led directly to the Ik's predicament.

As Western, and now Asian, tourists travel farther afield, and with the phenomenal growth of so-called ecotourism from the 1980s, the Third World is being increasingly called upon to devote land to national parks. Such land dedications are much easier in lightly populated countries like Namibia than in overcrowded ones like Rwanda. The 1992 Rio Earth Summit seems to have only exacerbated matters. Across the Third World, numerous Rio-inspired park creation schemes have been launched by the West with appalling consequences. In Ethiopia, the European Union (EU) proposes to force 7,000 people out of three national parks for conservation and tourism projects. Another EU-funded project in Uganda led to the forcible eviction of 35,000 people from the Kibale forest region. "Over three terrifying days, local people who resisted were shot or burnt alive in their homes, women were raped, their homes and crops torched. They had nowhere to go" (Harrison 1997, 23). I (Porteous) received anecdotal evidence about the creation of a park in highland Nepal. The former inhabitants were relocated to a lowland subtropical region to which they failed to adapt, suffering illness and premature deaths.

In Thailand, which has wastefully destroyed its own forests and now begun to work on those of Laos and Cambodia, the Royal Forestry Department is keen to double the nation's surface area of designated "conservation forest wilderness." This massive rezoning of forest land may eventually cause up to 1,000 communities to be resettled. A pilot project begun in 1994 attempted to resettle 2,357 people from the Doi Pha Chang wildlife sanctuary, most of them from ethnic Hmong and Mien hill tribes. Despite an array of land, housing, and monetary incentives, villagers were reluctant to move (Phatkul 1994, A3). Unfortunately they were up against not only the Royal Forestry Department but also the Protected and Prohibited Areas Committee, which includes the local governor, the provincial forestry officer, the district chief, and representatives of the social welfare office, the land reform office, the army, and the border police.

Less vulnerable than the Ik or the Hmong to the insistencies of profitable but thoughtless conservation-and-tourism policies are the Masai of East Africa. Unlike the Ik, they refuse to be consigned to extinction so that wildlife may survive. Indeed, before the era of conservation and control, African nomads generally enjoyed a stable relationship with non-human species. Traditional Masai migration routes cut across the areas designated for parkland in the Ngorongoro Crater and Serengeti areas of Tanzania. Unable to enforce the exclusion of the Masai, park planners were compelled to create a face-saving "conservation unit" between the crater and Serengeti National Park, wherein Masai cattle graze under the supposed supervision of conservation officers. The Masai, who have proved to be an attraction for tourists, are permitted to go no further than the lip of the crater from where they can see but not enter traditional grasslands far below. "The African crater came to an end when the Masai, who had enriched its life, were banished from it. Now that has been succeeded by the European crater, empty of man as an inhabitant, but seething with man as a day-tripper" (Marnham 1987, 30).

Lest we be tempted to deduce that most national park domicide problems are to be found in the Third World, we address the case of Canada. Since the establishment of Canada's first national park at Banff in 1885, the national park system has grown to thirty-nine parks, representing 2 per cent of the country's land mass. Various interest groups have attempted to sway park management policies over the years, and according to Dearden and Berg (1993), entrepreneurs dominated decision making until about 1970, when environmentalists enjoyed a brief period of dominance, themselves giving way to Aboriginal peoples in the late 1980s.

During the 1960s and 1970s, several national parks were created in the Atlantic provinces of Canada: Kochibouquac in New Brunswick, Kejimkujik and Cape Breton Highlands in Nova Scotia and Gros Morne and Terra Nova in Newfoundland. While most of the Terra Nova residents accepted resettlement, albeit reluctantly, park development inevitably displaced rural homes and a subsistence economy based on the natural resources of the area – hunting, trapping, collecting wood, and land ownership. This raised the question of how this loss of resources could be compensated in an equitable manner. Frequently, people could not identify the amount of wood they would need in the future because they had always cut wood as needed throughout the year. Replacing an old house on several acres of land with a new house and 1.5 acres of land may have appeared adequate, but it resulted in a considerable reduction in potential use. Felt (1977, 74) concludes that those who benefit are certainly different from those displaced, and to a large degree, benefit in the creation of parks is to a public interest not including the rural worker who previously inhabited these areas. This often occurs because planners assume that "the marginal rural worker does not particularly value his lifestyle, and that he is not being replaced in the next generation." In other words, the rural lifestyle is being phased out.

An example of a community that might have been displaced but escaped is Mountain Cove, Newfoundland. It was to have been resettled due to the creation of the Gros Morne Park in 1970. Here, again, there is evidence of an attachment to a way of life special to the home area. One resident suggested that resettlement would be particularly difficult for older people (Matthews 1976, 76–7): "On account people have their own home here, their own land, and their own fishing gear. You might say they have their own living here." When the persuasive government agents came to discuss resettlement, they were met with reproach by one of the community leaders and a response they probably did not expect. "If we build a house for you in Stephenville like you have here, would you move? I said, 'Not if it was a golden house would I move to Stephenville.'" According to Matthews, people did not believe that they would be forced to move because they felt that resettlement did not make sense. However expressed, their resistance was effective, and Mountain Cove was excluded from the park boundaries.

Aboriginal Canadians had not fared well against park establishment until 1979, when a new national parks policy acknowledged that future parks would be established only in conjunction with the settlement of native land claims. Confronting entrepreneurs and environmentalists alike, native peoples questioned the assumption that humans had no rights in parks other than as visitors (Sadler 1989). Regarding the land as ancestrally theirs, and perceiving themselves to be part of nature rather than separate from it, both Indians and Inuit began to take up an increasing role in park management, typically negotiating for co-management agreements and hunting rights within national parks.

Success has varied. Berg (1990) has investigated the relationship between park managers and the Nuu-chah-nulth Aboriginal people, who inhabit the area on the west coast of Vancouver Island where the Pacific Rim National Park reserve was established. The Nuu-chah-nulth people have been there for 4,000 years and the park for just over thirty years, but the natives have little say in park planning or in the management of traditional lands, their only concession being permission to continue the subsistence harvesting of seafood. In contrast, the Haida Nation to the north on the Queen Charlotlte Islands are guaranteed access to the Gwaii Haanas National Park Reserve for a host of traditional activities, including hunting, gathering, fishing, the cutting of trees for ceremonial or artistic purposes, and a variety of activities of traditional spiritual significance. Similar arrangements

have been made with the Mingan Band regarding the Mingan Archipelago National Park Reserve on the north shore of the Gulf of St.
Lawrence in Quebec. It is likely that future Canadian national park
designations will follow the Haida/Mingan model (Dearden and Berg
1993, 209).

In the 1980s, Canadian parks managers moved away from a view of
parks and protected areas as largely separate from other land uses to
one emphasizing coordination and integration on natural, social, economic, and broadly human ecological grounds. Put another way,
Canadian native peoples have forced the national parks bureaucracy to
abandon the American "empty wilderness" model and move closer to
the European notion of national parks as cultural landscapes – someone's home. This being so, there is no reason why non-native settlements pursuing traditional rural lifeways should not be included in
future North American national parks. The alternative – "pristine"
wildernesses swept clean of human content, balanced by fake Disneylands such as Williamsburg, Virginia – is suitable only for those so
divorced from the authentic that they can no longer distinguish
between shadow and substance.

Dams and Reservoirs

One of the most heart-rending images of domicide was relayed to the
world in July 1995. As the thirty-two billion litres of the reservoir
known as Lake Vagil, in Tuscany, were drained for dam maintenance,
a ruined village emerged from the depths. The village of Fabbriche di
Careggine was drowned half a century ago as heartbroken villagers
wept, some refusing to leave until water began to pour into their
beloved homes. Fifty years later, the survivors, mostly white-haired
elderly women clad in black, picked their way through the mud-caked
streets, lamenting. The village was theirs for a few days, before, once
more, the waters rose and their village was again lost to them.

In a sense they were fortunate, for of all forms of domicide, the
flooding of valleys behind dams is the most irreversible. Home is lost
at several scales at once: dwelling, neighbourhood, village, land, and
landscape are obliterated. The case of Lake Vagil is rare; most often,
all physical props to memory are drowned beneath the surface of a permanent engineered lake.

The numbers involved in dam building are staggering. Swift (1995,
8) estimates that in the mid-1990s the world contained 36,000 large-
scale dams, increasing at the rate of 170 per year, and displacing about
one million "reservoir refugees" annually. Some of these are super-
dams, defined as having a height of 150 metres and a reservoir with the

storage capacity of at least 25,000 x 10⁶ cubic metres. Super-dams are the largest structures ever built. They alter ecologies on an enormous scale, creating whole new geographies. In their rush to irrigate, they drown the most fertile lands, those at the valley bottoms.

Until the 1970s, the world's most active dam builders were located in the North – most notably the United States, the USSR, Canada, and Japan. Dams were seen as symbols of progress; as with the railways of the nineteenth century, folk songs were written about them. Except for Canada's James Bay project, large-scale dam building in the North ground to a halt in the 1980s, partly because of the potent effect of anti-dam activists, but also because there wasn't much left to dam. By 1990, the industrialized world was planning only thirteen super-dams, whereas in the South, at least eighty of them planned to tap the immense but relatively unexploited hydroelectric potential of the valleys of the developing world. Learning from the North, Third World leaders came to see large dams as symbols of progress, economic development, and independence. Since the 1950s, about 180 super-dams have been built in industrialized nations, but less than one hundred in the developing world. In the early twenty-first century, this imbalance may well be made up. The eighty super-dams on the drawing board for the Third World in 1990 included: twelve for Brazil, ten for Argentina, twelve for other parts of Latin America, ten for India, twelve for China, eleven for tiny Nepal, eleven in other parts of Asia and a mere three for Africa (currently the least dammed continent) (Lewis 1991, 36).

One of the world's least-dammed major rivers is the Mekong in China and Indochina. Current plans involve fifteen super-dams on the main stream and over 200 large dams on the tributaries. Not only would this plan fundamentally alter the ecology of several nations and threaten the livelihoods of sixty million people, but many valleys, with their fertile land, towns, and villages, would simply vanish from the map.

Further south, in Southeast Asia, the Kedung Ombo Dam in central Java has flooded the homes and farms of 30,000 peasants. When the floodgates closed in January 1989, hundreds of families living in the twenty villages to be inundated refused to move, protesting that they had not been consulted about the destruction of their property at any stage of the development. These families preferred to remain and cultivate the old flood plain as it re-emerged each year when water was released for irrigation (Probe International 1995). Next door, in Malaysian Sarawak, the Bakun Dam was begun in 1996. Ten thousand people from five ethnic groups were involuntarily resettled to an area thirty kilometres from the dam site (International Rivers Network

2001). As is usual in Southeast Asia, the project was cloaked in secrecy, and no social impact study was done. Hong (1996), who has studied the indigenous peoples of Sarawak, estimates that these people, mainly of the Penan, Kayan, and Kenyah groups, will lose their ancestral lands and probably their cultural identities. While the Asian currency crisis has caused deferment of the project, authorities are still considering construction of the project to its original size.

The indigenes have formed the Bakun Region People's Committee to fight legal battles in the Malaysian courts. They are caught, however, in the trap of the drive to development espoused by Malaysian elites, who have little but contempt for the people of the forest. If conditions in neighbouring Thailand are anything to go by, the indigenes will still be asking for compensation twenty years after the fact, holding demonstrations and being beaten and arrested by the police (*The Nation* (Bangkok), 23 December 1994). Indeed, the Penan have been harassed and imprisoned by Malaysia, and even Prince Charles has been roused to defend them (Sesser 1994, 244).

India has long been a leader of the Third World. It is also, with Brazil, one of the world's foremost dam building states. Some 1,500 large dams have been built in India, which presupposes many thousands of small ones; fourteen of this total are super-dams. Overall, dams account for 14 per cent of India's public expenditure. Yet, most of these dams gradually lose their value because of siltation. A study of fifty dams found that the average benefit accrued reached only 50 per cent of projections (Lewis 1991, 38). India's founding prime minister Nehru believed that an orgy of dam building would enable developing nations to catch up with the West, but this has proved to be in error. Water supply remains one of India's chief problems.

The domicidal effect of India's dam building is equally large in scale. It is estimated that at least eleven million (Lewis 1991, 38), but possibly as many as sixteen million (Pearce 1991, 20), people have been evicted in favour of reservoirs since the country was founded in 1947. Of these, at least half must still find new land to farm. Nevertheless, India has pressed ahead with the enormous Narmada River Project (NRP), a basin-wide development scheme along the 1,300 km Narmada River, which will harness it for irrigation, electricity generation, and domestic consumption through a series of construction projects reputed to require an investment of over $10 billion. The NRP involves building two super-dams, the Sardar Sarovar (SSP) toward the mouth of the river in Gujarat state, and the Narmada Sagar (NSP) complex upstream in Madhya Pradesh. In addition, thirty large dams, 130 medium ones, and 3,000 minor ones will be required. Such numbers are staggering.

Lewis (1991, 35) estimates that 3,500 square km of forest and 600 square km of productive land will be drowned, and that 1.5 million people will be displaced. Ruitenbeek and Cartier (1993, 13), consultants working for the project, claim that "over one million people" will be displaced. I (Porteous) was assured by an Indian official that it would be *only* about one-quarter of a million. These are enormous numbers of homes and livelihoods destroyed, and the potential for trauma and tragedy is vast. Typically, dam economists speak only about the budgetary requirements for compensation, which are estimated as one-time requirements of $275 million for the SSP and $425 million for the NSP. Such budgets are entirely inadequate, for the economic losses from displacement alone will be of the order of $100 – $500 million per year (Ruitenbeek and Cartier 1993, 13).

While proponents of the scheme laud its potential to irrigate, supply water, generate electricity, and fuel an industrial boom, opponents claim that the NRP will be a catastrophe environmentally, socially, and politically. In particular, the flooding of Narmada valley forests will flush out and destroy a rich variety of tribal cultures that have ancestrally predominated in the region – notably those of the Bhils, Bhilalas, Korkars, Gonds, Pardhans, Bharias, Bhumias, and Kols. According to Survival International, these people have nowhere else to go, and, instead of recreating their cultures elsewhere, they will end up in India's urban squatter settlements as yet more refugees from man-made disasters. This process has, in fact, already begun. Well-organized protests, although causing some foreign lenders to back out of the project, seem to have only caused NRP officials to harden in their determination to continue construction.

But the vast protest marches common in India are not permitted in totalitarian China, another of the world's major dam builders. In 1992, the World Bank approved a $120-million loan for the Shuikou Dam in eastern China, which is expected to displace 63,000 people. Of those already moved, "forced resettlement has radically changed their lives, resulting in loss of control and loss of face for the traditional farmers. In despair, some old people have committed suicide" (Probe International 1993a). The Erlan Dam is expected to displace 30,000 people, while the proposed Xiaolangdi Dam on the Yellow River will evict 190,000. "The resettlers are only being offered half the land they originally farmed, at a barren site, and without irrigation. If they want irrigation, they must dig their own channels; however, it is highly questionable whether they will be able to do so" (Probe International 1993b).

The "mother of all dam projects," however, is the well-known Three Gorges project, which involves building the largest single dam in the

world on the Yangtze River. This project will take about two decades
to complete, will cost perhaps $40 billion, and will require the reset-
tlement of over 1.3 million people. The reservoir created will stretch
over 600 km up the Yangtze, submerging 320 villages and 140 towns
and cities and some of China's most famous historic, cultural, and aesthetic landscapes. At least 1,200 historic sites will be drowned, and little is being done to rescue them (Higgins 1997, 22).

Besides claiming to provide all the usual benefits of power generation, flood control, irrigation, navigation, and the like, the Three
Gorges dam is clearly a politically prestigious project. The current
leaders of the people who built the Great Wall and of the world's most
populous nation see the project as a symbol of China's bid for world
power status in the twenty-first century. At the ceremony to celebrate
the diversion of the Yangtze, held in November 1997, the Chinese President Jiang Zemin, himself an engineer, claimed that "only socialism
can concentrate forces to create such big projects," while Premier Li
Peng, another engineer, was quoted as saying that the blockage of such
a great river "demonstrates to the world that the Chinese people have
the ability to build the biggest and most beneficial ... hydroelectric project in the world ... It not only inspires people, but demonstrates the
greatness of China's development" (Mickleburgh 1997, A8).

Criticism of the project was widespread before 1989, when the
Tiananmen Square massacre and the crackdown on dissent that
followed quickly silenced opposition. Such criticism is now severely
forbidden, and a book of critical essays edited by Dai Qing (1989, English edition 1994) was banned and pulped. Its information and opinions are available only abroad. Another volume *Damming the Three
Gorges*, aptly subtitled *What the Dam Builders Don't Want You to
Know* (Barber and Ryder 1990), condemns the project on a wide
variety of grounds.

In terms of resettlement, the plans call for a system known as "resettlement with development," which is based on consultation with the
affected population as well as popular acceptance by the "host population" in the resettlement areas. Master planning would ensure that
those resettled would be provided either with adequate farmland or
jobs equivalent to those they were to lose. But the plans also contain
clauses that restrict the right to migrate to cities, while containing no
provision for the estimated 10 to 30 per cent of the drowned towns'
populations who are illegal migrants. This considerable floating population is simply regarded as non-persons. By 1991, 50,000 people had
already been uprooted in pilot relocation experiments (Adams 1993,
19), but Fearnside (1993, 57) found little evidence that any of the local
people had been widely consulted. Further, there is simply no land

available that is equivalent in quality to the land lost. And, finally, China's previous resettlement projects have been plagued by "mistakes" such as waste, mismanagement, and inadequate funding.

China's leaders still appear to have a Great Leap Forward mentality, oblivious of the fact that thirty to forty million people died of famine during that period. They are "afraid to lose face," asserts one dissident professor (in Higgins 1997, 22). Their moral bankruptcy is seen in their pressure on the Hong Kong Film Festival in 1991 to withdraw the film *In Expectation*, which deals with village extinguishment, with the massive corruption that has dissipated resettlement funds (Mickleburgh 1998, A9), and with the subtext of the plan. This is part of the move to further centralize power in Beijing (Sullivan 1995, 269; Zich 1997, 23).

Meanwhile, Western visitors bring back evidence of the anger of the relocatees: "Are we going to have enough land to farm? No ... There is no land to farm behind our village. When the time comes, I will refuse to move out of my village. If they want me to leave, they will have to use police to drag me away" (Mickleburgh 1998, A9). "No good! I was born here. I built this house with my own hands. I'll lose my inn! For me, it's terrible" (Zich 1997, 28). Other interviewees displayed the usual array of emotions in such situations – from anger and sadness to resignation, acceptance, and even anticipation of a better life. Few felt that anything could be done to oppose the process or even to prevent the embezzlement of resettlement funds by officials: "I'm only *laobaixing* [of the common people]. Officials don't care about the *laobaixing*" (Zich 1997, 23). Oblivious of this inhumanity, Western tourists take scenic cruises along the Yangtze, having heeded the call, "Go now before it's too late!"

While it is impossible to fully judge the domicidal effects of the projects outlined above, most of which are currently in progress, a review of several already completed projects may provide an indication of what the twenty-first century has in store for Brazilians, Indians, Chinese, and other likely victims of dam building. The Norris Basin of the Tennessee Valley Authority, the Volta River Project (Ghana), the Kariba Gorge Dam (Zambia), and the Aswan Dam (Egypt) all involved comprehensive planning and promised economic renewal, but in each case they caused the end of a way of life for those whose homes were drowned. Two British Columbia projects (the W.A.C. Bennett Dam and the Kemano project) met the needs of industry and power planning but not those of the First Nations peoples who were displaced.

The first example comes from the Great Depression years in the United States, which saw the development of many projects by the Tennessee Valley Authority. This Norris Basin project included the introduction of land use planning, regional development, and multi-purpose stream development. But it also involved land purchase, family removal, and relocation, albeit from an area that was isolated and economically disadvantaged. The project had a proponent in a position of ultimate authority. Franklin Delano Roosevelt stated in *On Our Way*: "Before I came to Washington, I decided that for many reasons, the Tennessee Valley – in other words, all of the watershed of the Tennessee River and its tributaries – would provide an ideal land use experiment on a regional scale embracing many states" (McDonald and Muldowny 1982).

Through a series of oral interviews, McDonald and Muldowny have created a picture of the basin before it was flooded and also a sense of the positive and negative aspects that this development brought to people who lost their homes and their way of life as part of Roosevelt's "ideal land use experiment":

And when they were leaving here, it was just like a funeral here at the store every day – they didn't know whether they'd see them any more in their life.

It's very difficult to describe the attachments that they had for their land, their emotional involvement, and the fact that they were going to have to leave all that and come somewhere else.

It wasn't just that they had spent all their lives there, you know, but as far back as their grandparents could remember.

Well, I didn't feel too awfully bad about it. One way of looking at it, I just thought that when the government took a notion to do anything, they just done it, and I just passed it by, is all I can say.

I don't know how – they *existed*, they didn't *live*. Now, [for] the people like that, I guess in the end it was better – turned out that it was better for everybody that they did move.

I know some people up there that resented moving at all. In fact, I know two who committed suicide. I knew personally of them. They bought all around [one man], but he wouldn't sell, and he went down to the pond there and put a rope around his neck and hung himself.

The thing that hurt so bad was that we just didn't want to be taken away from the place we loved. Even if we went away, we would like to come back and see the place again. Now, it's a hundred feet under water. We can never go home again.

These interviews provide us with an opportunity to listen to the feelings of some of the persons who lost their homes through the creation of a reservoir. In the next three examples (in Ghana, Zambia, and Egypt), the commentators have provided a critical view of the process by which domicide occurs as well as some insight into people's reactions.

The development of the Volta River Project in Ghana in the 1960s necessitated the flooding of an area of 8484 square kilometres. Before the water submerged traditional lands and homes, some 80,000 people lived in the area. This was a project for which careful plans were made. The British and Gold Coast governments announced the establishment of a Preparatory Commission in 1952 to report on technical aspects and economic viability. All aspects of project construction were studied, including human issues such as compensation and resettlement. But, in retrospect, the chair of the commission said: "Those of us who are not Ghanaian ... clearly realised how little we understood the minds of the people who would have to leave their ancestral lands and homes"(Chambers 1970, 5). Land had particular significance for the Ghanaians. Traditional beliefs stressed the connection between land and life (Amarteifo 1970, 131).

Problems arose due to the long delay in obtaining financing for the dam and internal difficulties with staffing the secretariat for the project from existing government departments. Then, dam construction was suddenly begun, and there were only three years to remove the residents of the basin. Further tensions arose in trying to implement self-help policies while still meeting the minimum housing standards set by the authorities. This caused costs to rise significantly. Villagers who were to be given the opportunity to construct their own dwellings lost enthusiasm when considerable speed was necessary to accommodate the coming flood. Eventually, the Department of Social Welfare recommended that housing projects should be built, and that self-help projects should follow as "an excellent way of rehabilitating [the victims] and giving them a stake in their new environment" (Kalitsi 1970, 40). Reviewing the limitations of planning in this project, Huszar (1970, 161) notes that while planners can create the physical manifestation of a new settlement, they cannot create the same social and economic environment. Thus, persons who were resettled lost their traditional house forms and occupations.

Resettlement of the Gwembe Tonga in Zambia in the 1960s was brought about by the construction of another African project, a large hydroelectric dam across the Zambezi River at Kariba Gorge. The social consequences of this resettlement as described by Colson (1971, 2), included hostility toward government, loss of legitimacy of local leaders, and questioning of existing religious practices. On the positive side, Colson reports a new focus on kinship ties. Colson suggests that resistance to technological change was experienced not because the Gwembe Tonga were averse to the proposed change, but because the resettlement program threatened people's basic securities such as land for cropping and grazing; those resettled did not understand the technical facts, the project resulted from a command from outside their community, and the Gwembe Tonga believed that they were made to suffer for the longer-term good of their community. While new settlements were eventually created, that which was lost is expressed in a rather special way for the Gwembe Tonga: "The resettled people came from an area of old settlement, where all the features of the landscape were named and people were easily oriented in space. They came from a landscape threaded with paths linking one homestead with others in village and neighbourhood, homesteads with fields, fields with one another, neighbourhoods with other neighbourhoods. In most resettlement areas, there were few game tracks, and while newcomers recognized major landmarks if they were lucky enough to move into the uplands, most of their new world was anonymous" (Colson 1971, 50).

One final example in Africa involves the Aswan Dam. The Aswan Dam was completed in 1970 and created the world's third largest reservoir, Lake Nasser. The reservoir was named after Gamal Abdel Nasser who was then president of Egypt, advocate of the dam's construction, and a major influence in the securing of funds for construction. While the dam benefits Egypt by controlling the annual floods, providing about twenty percent of Egypt's power supply, and improving navigation, problems have developed through seepage, evaporation and siltation. The costs to those displaced were also high.

Ultimately, this project was responsible for the displacement of 100,000 Nubians in Egypt and the Sudan. Hussein Fahim undertook research in Egypt in 1969, 1973, and 1980 to study the means of coping developed by those who had been displaced. His account was prepared to assist persons responsible for future resettlement programs to understand "the instinctive need for mankind to preserve individual cultures"(Fahim 1983, x). The construction of the High Aswan Dam displaced all Nubians within Egypt and flooded one-third of the Nubian lands in the Sudan. While the Egyptian Nubians were resettled within the Aswan area, the Sudanese Nubians were located in a unfamiliar area. It has been suggested that a pattern of out-migration had long been set for the Nubian people before construction of the dam. But many Nubians had expected, despite the poverty of the area, to return from the big cities to retire, to enjoy a life of freedom and independence, to a country of silence and beauty, and to houses and extended compounds that had been in the same tribe for generations. With the dam construction, some Nubians became, in their own words, "inflicted people" who suffered from depression and grief.

This displacement occurred despite the fact that the Egyptian government took the problem of resettlement seriously. While the Nubians were not given an active role in the formulation of plans, they were consulted, and a compensation policy was developed that would be suitable to their needs. At first, the Nubians did not really believe that their valley would be drowned, but when the time came for resettlement, some saw it as an opportunity for new material and social advantages. Others saw an end to their peaceful and quiet way of life. Fahim (1983, 43) tells of the day of departure for the first village when "the women rose at dawn to sadly and silently visit their dead, spraying the graves with water, expressing compassion and sanctification." Many "kissed the land as they left their empty vacated homes, while others filled their pockets with small bags of soil." Every effort was made by government officials to welcome them to their new location, and new houses were ready for them. Unfortunately, even with this excellent start, problems surfaced thereafter including delays in scheduled moves and in the provision of new housing.

Many of the efforts to resettle persons displaced by the creation of reservoirs in Africa were missing in two contemporary British Columbia projects. The W.A.C. Bennett Dam created Williston Lake in the late 1960s by flooding the Peace River along 632 km of the Rocky Mountain Trench, thus drowning a valley of historical significance that was the homeland of the valley's residents – the Sekani. Some of the Sekani refused to leave, and on 25 May 1968, the provincial government sent in bulldozers and pulled their houses to higher land. Living in poverty, due to the flooding of traditional lands, and suffering health problems since then, some of the Ingenika people of the Sekani have been reluctant to leave Ingenika Point, which is the only part of their traditional lands remaining. One woman said: "I'd get heartbroken ... to move from a place where we've been so long ... when I go away, I want to come back fast ... I get real lonely ... It's a nice feeling to get back home" (Cruickshank 1987, 1). Only in 1987 did the British Columbia government recognize the plight of the Sekani and begin to provide aid to their settlements. While the Bennett Dam and the Williston Reservoir glorified two local politicians and became the primary source of electrical power for British Columbia, they diminished an already-suffering Indian nation and became a primary source of sorrow and suffering for the Sekani (Jensen 1996, 7). Investigations by newspaper reporters (Glavin 1987) and scholars (Koyl 1992) confirmed that the native people had been disregarded, marginalized and mistreated.

White people in the area, primarily in the community of Hudson's Hope, were treated with greater courtesy by government officials and the BC Hydro corporation. Nevertheless, local residents were sufficiently disturbed to publish their own book, *This Was Our Valley* (Pollon and Matheson 1989). In it, they describe the expropriation and compensation process as follows: "They [BC Hydro agents] sniffed through the community, smelling out its most avaricious members, those most susceptible to an offer. They spread rumours; they spread lies ... they played on the social conscience of community members accusing them of selfishness, of denying the greater good to the greater number. And, in the final resort, judiciously at first, then more threateningly ... they invoked the prospect of eminent domain. They did this all without a sense of shame because they told themselves they were serving an ultimate good" (209, 212). In Hudson's Hope, white people lost some homes. Economically, however, the town gained considerably. Not so the Dene Ingenika, who developed both physical illnesses and severe symptoms of psychological stress, "so we don't know which is the real sickness, or which is in our minds. But it didn't used to be" (Isaac, quoted in Pollon and Matheson 1989, 339).

Another British Columbia reservoir created by the Kemano Dam on the Nechako River severely affected the Cheslatta Indian Band. The treatment of the Cheslatta Carrier Indians has been described by a member of the federal Royal Commission on Aboriginals as "a story of horrors" (Hume 1993, Robertson 1993, 1; Wagg 1993, 1). The Aluminum Corporation of Canada (Alcan) called the Department of Indian Affairs in March 1952, saying that the Cheslatta Band would need to be evacuated. According to Marvin Charlie, who later was elected chief of the Cheslatta people, they were self-sufficient and knew nothing of the government until 3 April 1952 (Charlie, 1993). On that day, the Department of Indian Affairs told the band to move from their homes immediately. On 8 April the gates of the dam were closed and the water started to rise, flooding some buildings. The Indian agent called a meeting for 16 April, and the people gathered, but neither Alcan nor the Department of Indian Affairs arrived. Finally, on 21 April, when many people had run out of food, officials arrived and started to negotiate for land. Since there was no native chief, the Indian agent appointed a government chief and two councillors. Few Indians spoke English.

People asked for compensation for "life" because they said that the dam would destroy their way of life. Alcan promised that they would

replace what was left behind, pay for the land, and move two graveyards. Some graveyards were too old to move, and so the grave houses and markers were gathered together and burned. A sign was placed
on them, which, as Chief Charlie recounts, reads: "Here be mens,
womens, children. May they rest in peace." His description of this sign
is somehow more affecting than the reality, which is an aluminum
plaque reading: "This monument was erected in 1952 to the memory
of Indian men, women, and children of the Cheslatta Band, laid to rest
in the cemetery on Reservation Five, now under water. May they rest
in peace."

The negotiations took two days. Some elders refused to move but
were told that the law would come, and then there would be no compensation. Often surrender was executed by taking each individual out
of the meeting separately. The first person would be offered $5,000,
the next $6,000. Many of the younger people left on saddle horses with
only what they could carry. Two weeks after the move, all the houses
were burned, barns were torched, and everything was plowed over by
machines. The Alcan contractors refused to burn the church, and in
June, the Indian agent flew in by helicopter and burned it down. Charlie says that, despite having to wander around for three years to find
land to buy: "All this did not bother us so much ... but when [the]
graveyard was flooded [including the grave of one of his brothers], that
really hurt my people."

The Cheslatta people moved to the Grassy Plains area and lived in
tents and cabins. "We were refugees in our own country" (Charlie,
1993). They bought their own land, and despite promises, have never
been reimbursed for their lost land or equipment. They lost their livestock, their way of life, their way of hunting, and almost their own language. When Charlie was elected chief in 1990 he noted that in the old
graveyards, not one person buried there had died of alcohol or suicide.
In the new reserve, there were three huge graveyards and every other
death was from drug or alcohol abuse. Ninety-seven per cent of his
people were on welfare. This factor simply exacerbated the effects of
loss of home, all of which have been redressed in part by Marvin Charlie. He has worked to create job opportunities for the young people on
his reserve, and now the welfare roll has dropped to 35 per cent. They
have "found out where their roots were, where their ancestors were ...
bringing them back to the ground where they come from." All this was
again threatened by the Kemano Completion Project, which, through
lowering of water levels on the Nechako River to create more hydroelectric power, would, according to Charlie, ensure that "the lake is
going to die, and my people will die too." The British Columbia government's announcement on 23 January 1995 that this project had

been cancelled may have come as good news to the Cheslatta; concern about protection of the fisheries resource remains.

The disregard for native peoples in the creation of major projects was summed up by Isak Dinesen thus: "It is more than the land you take away from the people, whose native land you take. It is their past as well, their roots and their identity. If you take away the things that they have been used to see, and will be expecting to see, you may, in a way, as well take their eyes" (1972, 375). Yet, our final example of domicide, the Oldman River Dam in Alberta, points out the similarities between people who use the land for their livelihood, whether they are indigenous or settlers. Million's doctoral dissertation, *It was Home*, takes its title from a dispossessed white rancher, whose words surely form an epitaph for all such circumstances: "It was home. So thinking of leaving it was painful. But thinking it was to be all gone, all destroyed, that was the hardest part. I mean you'll never be able to see it again, to walk on it again" (1992, 131).

Million's work on the Oldman River Dam ranch displacements is useful in its consideration of the cycle of attachment, uprooting, and resettlement. The founding of place involves "work-in, pleasure-in, name-in, and living-within-place," while the subsequent growth of belonging includes "everything-in-place, habit-in-place, and time-in-place." The process of uprooting and forced journeying demands a sequence of "becoming uneasy, struggling to stay, having to accept, securing a settlement, and searching for the new," while the final stage, rebuilding place, involves "starting over, unsettling reminders, and wanting to settle."

Further useful insights to be found in Million's work include: the close "fit" that emerges between individuals, their homes, and surroundings within the geography of a particular place; the pleasures of working day in and day out on a place, often a place that holds a family's history; the existence of place as a totality of habits, and the deepening of care that occurs for a place; the moments of "being uneasy" and "seeing to believe" as projects move from the drawing board to reality, followed by the "struggle to stay"; the belief that confrontation is hopeless, that displacement is a contribution to the "common good"; that being unable to return does not preclude recollection of place by way of memory, but this past may be regarded as make-believe; and that having to struggle to secure compensation causes an additional violation of place, and that compensation principles overlook the "pragmatic working totality" and the "embodiment of identity" of place (Million 1992, 133). Million's study provides an opportunity to learn directly from individuals about the process of being displaced, which materially assists our understanding of domicide.

In sum, the economic advantages of dam building are well known and, in some cases, worthwhile. But there are also less obvious advantages that accrue to engineering companies, development consultants, United Nations agencies, major financiers such as the World Bank and the Asian Development Bank, rich governments with aid to distribute, and even those hired to supervise domicide with the least uproar. While many of these advantages are indeed economic, they are also social and political, centred on the immense prestige that accrues to those engaged in megaprojects. Profit and prestige are quite sufficient for most of those involved to overlook, minimize, or explain away the tremendous environmental and social effects of such dams (Goldsmith and Hildyard 1984).

Nevertheless, international opposition to both the environmental outcomes and the plight of the one million "reservoir refugees" uprooted each year has been vociferous. From the Sommers' (1984) quirky *Scenic Drowning* to Cummings's (1990) *Dam the Rivers, Damn the People*, many publications now chronicle the development of resistance not only among international observers but, increasingly, among those victimized by dam construction. This resistance is best chronicled later in this book; we leave this section with a thought from that Cold War novel of almost half a century ago, *The Ugly American*: "We finance dams where the greatest immediate need is a portable pump" (Lederer and Burdick 1958, 238).

CONCLUSION

The central conflict of the twentieth century [is] the efforts of individuals, families, and communities to preserve their freedom against the over- whelming power of the techno-industrial superstate.

Edward Abbey (in Balian 1985, 59)

This review of domicide that occurs on an everyday basis through urban redevelopment, economic restructuring, or the placement of public facilities compels us to revise somewhat the conclusions gained from the earlier study of extreme domicide. Here, the process is much more subtle. Motives, for the most part, may be legitimate or even admirable. Common good rhetoric makes more sense to the victims of everyday domicide; even removals for purely profit-making enterprises can be cloaked in the language of public interest, such as the genera- tion of more local employment. Further, everyday domicidal actions are almost invariably legal. These characteristics contrast with those of extreme domicide, in which case subtlety is often absent and motives, legalities, and common good rhetoric are usually suspect.

Further, incidents of everyday domicide tend to occur on a smaller scale, although super-dam projects can affect whole regions. And everyday domicide is distinguished by its frequency; the chances of being caught up in it can be quite high, depending upon who you are and where you live. Governments are the proponents of many projects that cause everyday domicide; their minimal role is to provide regulation when projects are proposed by private interests. Leadership may come from powerful politicians with particular visions of progress, such as Franklin Delano Roosevelt in the United States or W.A.C. Bennett in British Columbia. Other decision makers include high-level bureaucrats who closely identify with the success of their projects and have little interest in the communities they will inevitably dismember or expunge. The benefits accrued by such persons are not primarily economic but psychological – the sense of esteem that comes from the successful implementation of a major project.

Significant planning activity often occurs before and during the projects, including, in cases such as the Roskill Commission, several years of elaborate inquiry. Despite this, the values of those to be displaced are often not adequately recognized. In Third World cases, in particular, traditional cultural relationships to land are generally ignored. Inadequate communication between planners and planned-for is, sadly, almost commonplace. Too frequently, those most affected simply have no effective way to make their wishes known.

Protest against the prospect of domicide does occur, but it is not often effective unless there is clever use of the media, as was the case with the proposed third London airport. Usually, however, people believe that confrontation is hopeless. During project implementation, the expropriation process is often inadequate, leaving people to suffer through anxiety. Too often, there is evidence less of maltreatment than of lack of consideration for the victims. People forcibly relocated lose more than loved homes; they also lose social networks and a sense of belonging to both community and the physical environment that supports it. Many suffer anger, sadness, and grief, or even more serious physiological and psychological effects.

Finally, while extreme domicide was shown to be, in some cases, reversible, and certainly avoidable, everyday domicide occurs so often that those who are not involved treat it as part of normal life. It is neither avoidable nor reversible. While bombed cities are rebuilt and native peoples are recovering some of their ancestral lands, few of those drowned out by dams or concreted out by airports will ever have the chance to restore their homes. The following chapter presents two in-depth studies of this irreversible drowning of home.

CHAPTER FIVE

Drowning Home:
The Columbia River Basin
in British Columbia

Can we not maintain at least some semblance of dignity, some shadow of our former selves? By leaving now, by collapsing and admitting defeat, we only aid and abet the destructive plans of our occupiers. Should we not stay here as long as we can, and live what remains of our cherished lives here as fully and richly as we can? Each family that leaves now tears a permanent hole in the web of our community life. No new neighbours will come to replace those we lost: we are the last people who will live here, and we must band together. Let us leave only when we must. Let us leave together, at the end – not piecemeal, in panic and terror, at the beginning.

Frank B. Auberon, Sr., "Letters to the Editor,"
Paradise Valley Daily Transcript, (6 July 1927),
in Barrett (1992, 123).

In this chapter, we deepen our understanding of the concept of domicide. All that has been discussed in previous chapters is tested by exploring two British Columbia situations in which the drowning of home occurred as a result of the construction of reservoirs for the Keenleyside (High Arrow) Dam and the Libby Dam in the 1960s and in which, today, a new association has arisen based on that original drowning of home. This chapter asks the following questions: What did home mean to the area residents? What did the residents of this area believe they would lose when they lost their homes? What was the process by which they lost their homes? What were the motives behind dam construction, and who was responsible for these motives? Who benefited from dam construction? What was the reaction of people to losing their homes, and what effect did it have on them? and What plans were made to assist people who were to lose their homes?

In answering these questions, it will become evident that the construction of large hydroelectric projects provides a particularly vivid and significant example of home destruction. There is a finality to the destruction; it is not possible to rebuild or even view the landscape on which homes once existed.

Three major published sources deal with the Columbia River dams. The first, Waterfield's *Continental Waterboy* (1970), is an entertaining "tale of what happens to people, to their homes, and to their habitable valley when they get in the way of the apostles of progress and of hydroelectric engineering." Neil Swainson's *Conflict Over the Columbia* (1979), prepared from a political science perspective, clarifies the background to, and considerations involved in, negotiations of the *Columbia River Treaty*. The most significant source of information on the measures taken to assist people who were displaced by the Arrow Lakes project is *People in the Way* (1973) by J.W. Wilson. Wilson was responsible for the resettlement program on behalf of the project developer, BC Hydro and Power Authority. He tells "the story of the attempt to deal with the problems of human settlement and displacement resulting from the Columbia River project" and how frequently "the best laid plans ... gang aft agley" (Wilson 1973, xiii).

This chapter augments these published sources in its focus on hearing those who would lose their homes from the perspective of the process they endured and, in addition, bringing their story up to date. Following a brief general description of the years leading up to approval of the projects, public hearings relating to the projects are described. The meaning of loss of home and reaction to the process of losing home are discussed based on evidence from the public hearings and other available sources. Finally, the question of who benefits in these circumstances is discussed.

PRELUDE TO DOMICIDE – YEARS OF UNCERTAINTY

The Libby and Keenleyside Dams are part of a comprehensive system for flood control, navigation, and hydroelectric power in the Columbia River Basin that straddles the border between the Canadian province of British Columbia and the American states of Washington, Idaho, and Montana. Creation of the dams had been discussed by the United States and Canada from 1944, when the Canadian and American governments requested their International Joint Commission to determine whether greater use could be made of the Columbia River system. On 9 March 1945, Prime Minister Mackenzie King announced that the Canadian government had requested the commission's Columbia River Engineering Board to survey hydroelectric and flood control potential. Despite the presence of surveyors in the area from this time and warnings from their MP, H.W. Herridge, most

people took little interest in the proposal beyond their attendance at a
few local meetings. This changed when US Senator Richard L. Neuberger made a grand tour of the area and, in addressing the Nakusp
Chamber of Commerce, ended with the threat: "If you Canadians
continue to delay the building of storage, we shall have to consider
your behaviour an unfriendly act between nations." The dams were
represented by Neuberger as an essential response to the needs of a
growing population and to the improved post-Second World War
economy, thereby introducing the public interest argument to this
situation for the first time.

However, the need for the dams and for flood control through the
creation of upstream water storage suddenly became even more press-
ing. As a result of serious flooding in 1948, Trail, British Columbia,
was inundated, and in the United States, fifty lives were lost and $100
million in property was damaged. Recognizing the possibilities of comprehensive development including the provision of flood control, a
number of options were presented, and discussion relating to these
options ensued for many years.

In December 1960, Waterfield (1970, 35) reports that the Arrow
Lakes residents began to hear rumours from afar that an agreement
had been finalized between the United States and Canada. The
Columbia River Treaty was signed on 17 January 1961 and ratified
16 September 1964 following a bitter debate by a large section of the
Canadian population regarding the sovereignty of Canada's water.
The treaty's preamble consisted of the following words: "Recognizing that their peoples have, for many generations, lived together and
cooperated with one another in many aspects of their national enterprises for the greater wealth and happiness of their respective
nations." It required that Canada provide more than 19 million
cubic metres of storage by the construction of three dams: two on
the Columbia River – the Mica and High Arrow (Keenleyside); and
one on the Duncan River (Canada 1964). The British Columbia
Power Commission (subsequently known as BC Hydro) was named
as the Canadian entity responsible for constructing and operating
the three treaty dams. The Libby Dam was also accepted, and the
American Bonneville Power Administration was made responsible
for its operation. Of the four dams, the Keenleyside and Libby Dams
were the only two that significantly affected the population in
British Columbia.

In the words of one of the treaty's opponents: "The present treaty
assigns all the risks and obligations to Canada and most of the
benefits to the United States. An outsider reading the terms of the
treaty could be excused for assuming they had been imposed on a

conquered country which has surrendered unconditionally after a war. The terms represent a perpetual erosion of Canada's self respect; a national characteristic that already is in a frail condition" (Ripley 1964, 60). Before construction could occur, however, there were many years of uncertainty and pain for the perople who would be displaced. This had started years before, when public hearings (of a sort) were held.

PUBLIC HEARINGS

Many of the benefits of building the Libby Dam were confirmed by the International Columbia River Engineering Board and acknowledged by residents of the Kootenay Valley near Creston in the hearings that were held in July 1948 by the International Joint Commission (Swainson 1979, 43). Public hearings were also held by the International Joint Commission in Spokane, Nelson, and Cranbrook in March 1951. In fact, the hearings' transcripts reflect mainly the positions held by the provincial and federal governments and only to a much lesser degree, those of individuals in the area. The commission was represented by General A.G.L. McNaughton, chairman of the Canadian Section, and A.O. Stanley, chairman of the US Section.

The official statement from the Province of British Columbia did not oppose the project but requested prompt protection and indemnity of all interests in British Columbia that would be affected by the erection and operation of the works; recognition for loss of taxes and for the loss to the economy of the province of the productive value of the lands to be flooded; and an amount of electrical power to be delivered in British Columbia as the commission deemed appropriate.

The Government of Canada did not oppose the application either, citing its "sensitivity to the potential costs to Canada as a whole of a major frustration of a strongly-held American desire" (Swainson 1979, 46), but submitted that the approval should be subject to conditions to ensure protection and indemnity against injury of all interests in Canada that might be affected by the construction and operation of the dam and reservoir, as provided by Article VIII of the *Boundary Waters Treaty*, 1909, and a fair recompense to Canada for the utilization by the project of Canadian natural resources.

Of the few other statements from those appearing at the hearings, one is of particular interest. It is therefore produced at length. The statement to Commissioners A.O. Stanley and General McNaughton is by Jack Aye, a Canadian rancher from the South Country (Lake Kookanusa area), who was to lose his home (International Joint Commission 1951):

Aye: There are a few questions that I would like to ask on behalf of some of the ranchers in the district that are going to be flooded. At first we had thought about organizing, but on giving it further thought we decided that it would be futile for us to try and stop the building of this dam. I know myself for one, I am certainly not in favour of it, probably for a very selfish reason. We have spent the most productive part of our lives building up a ranch that will be flooded in this area, and the place is just now starting to produce. It is not into production yet, and it is not a very nice feeling to have something like this come on you. But, I do not feel there is anything I can do to stop it, but I would like to know how much longer we are going to have it hanging over our heads here.

If they are going to go on and build this dam I feel that we should get a settlement so that we can get out of there, and get started again before we are too late altogether.

I am afraid that I do not agree with Mr. Melrose here that there is plenty of room on land in the district to move us ranchers to. I have not seen any place in there, and I have lived there all my life, that would be up out of the flood, out of the river bottom that will be flooded that would take the place of what I have.

Stanley: In making this, just like when a railroad goes into a piece of land, or a mill floods it, it depends upon the value of the land, and the value by reason of what is destroyed by the loss of the land will be taken into consideration. All these things will be taken into consideration, and I do wish to assure you, my friend, that you would not have to wait until the dam is built to be compensated. That can be done before that, and it often is. You are worse scared than hurt. We are going to take care of you.

Aye: Yes, but there is another point there. It is very easy for a man on the outside to say, "Well, you are getting paid for it, you will get paid for it." How much? When? Where? If a man comes along to buy your place, you make a deal, and if you don't like it, you don't deal. This is a different proposition. You, gentlemen, the Government, they say: "Here now, we are going to build this dam." But I cannot say that I am going to stay there.

Stanley: Going to what?

Aye: I cannot say I am going to stay there. It is a very different proposition. It is a different setup altogether.

Stanley: These things happen every day. Every time you build a railroad, every time you build a highway from your market, from your farm to your market, you exercise, the Government exercises what they call the power of expropriation. It is the duty of every good citizen to waive his gains ... But I do believe in the right of a Government to do these things necessary for the benefit of all of us, and then [give] our good citizens ... complete and adequate monetary compensation, and that without delay.

Gen. McNaughton: I would like to state from the point of view of the Canadian Section of this Commission, I would like to reaffirm what my colleague has said, the Chairman of the American Section, that the Treaty provides that the payments for compensation have got to be on such a scale that the Commission are satisfied as to the equity, and so we are not only to help forward the project which needs to be undertaken in the general public interest, but we are here as trustees over and above the governments, in this case to measure the compensation to be paid to individuals who are hurt or damaged in the process of making their land and property available for the public good, that they will receive compensation which is just. Otherwise, no permission to build a dam can go. That is settled by Treaty. I can give you assurance on that score.

Stanley: In addition to that, the treaty provides that we must give you adequate compensation, and every individual must be compensated for his land or property at the time that it is taken. We will compensate you at the time that we take your property, and give you an adequate return for every nickel that it cost you. We do not want to split hairs with you over a shirt tail full of land.

Attendance at the second Canadian public hearing held at Nelson in March 1951 was similar to the Cranbrook meeting. Of particular interest is the statement by Guy Constable in favour of the application on behalf of the Associated Drainage District of Creston, the Associated Boards of Trade, and the Associated Drainage Districts. He noted that the project would allow the additional reclamation of 6880 hectares of land, which would no longer be subject to periodic inundation. This was equal to the amount that had been reclaimed up to that time – a clear statement of an expected benefit to the region, if not to all the people of the South Country.

The treaty enabling the Hugh Keenleyside Dam, which affected the Arrow Lakes area, was signed without benefit of any real representation from the people who were to be displaced. When, on 6 February 1961, a delegation from the area to be affected by the dam went to Victoria to protest, they are purported to have been told by Dr. Hugh Keenleyside of BC Hydro "that they should have faith in the people running the country and should accept their decision as being in the best interests of all concerned" (Canada, House of Commons 1964, 602).

The only hearings occurred in Revelstoke, Nakusp, Castlegar, and Victoria in the fall of 1961, and these were part of the water licensing process, which is primarily concerned with the impact of water use on the holders of existing water licences (Comptroller of Water Rights 1961 *a,b,c*). Unlike some jurisdictions that base water law on riparian rights, proprietary rights to water are vested in the Province of British Columbia, and the opportunity to use or divert water is subject to a licence. The issuance of licences is based on whether or not sufficient water exists. Consideration of intangibles, such as the significance of loss of home, was beyond the scope of the comptroller's hearings. In the words of one resident, the hearings were "a farce" (Wilson and Conn 1983, 46).

The editorial from the *Arrow Lakes News* (1961), a broadsheet published by BC Hydro for local residents, for the week of the hearings stated:

On Friday and Saturday of last week, in the Legion Hall, Nakusp, the town's residents and those from the neighbouring districts, Galena Bay to Edgewood, were "on the stand" to defend their homes and valley, from destruction, by the High Arrow Dam.

The "court" was conducted according to law, and we can well be proud of the very fine material presented by a large number of our citizens and the excellent manner of presentation. Dignity and sincerity were evident in most briefs, from those who spoke without notes to the most elaborately prepared document by the larger organizations.

The trust and considered thought put into all material presented made a favourable impression on the chairman, Mr. Paget, and also on Dr. Keenleyside, the chairman of the B.C. Power Commission.

No political prejudices entered the appeals. No superficial supposed reasons were offered. This was a "court" where each individual was fighting for that which was closest to him or her – his home, his living, and his country ...

Somehow, we feel that these expressions will not go unheeded, but will find a sympathetic ear, and a second look will be taken at the whole situation of Columbia development, to make it really development, not destruction.

But the Arrow Lakes' residents who spoke against dam construction were not successful, and the water licence appurtenant to the project was issued on 16 April 1962, the comptroller "being satisfied that no person's rights are injuriously affected" (British Columbia Department of Lands, Forests, and Water Resources 1962).

In the terms of the licence, the only clause that directly related to the human inhabitants of the reservoir area read that: "The licensee shall review with the Comptroller of Water Rights prior to expropriation under the Water Act or any other Act any matter where the licensee is unable to reach agreement with the owner or owners of land affected by the works and the operation thereof as authorized under the licence." Even after issuance of the licence, the Arrow Lakes residents continued to doubt their impending fate due to continued bickering between Premier Bennett of British Columbia and Davie Fulton, the federal minister of Justice. However, their sense of security was to be short-lived.

In April 1963, the Liberals came into power at the federal level with a narrow majority and signed an agreement with British Columbia that made it clear that High Arrow would go ahead. Then the federal External Affairs Committee was convened by the minister of External Affairs, Paul Martin, to accept or reject the treaty; no alterations were permitted. The treaty was accepted, the way was paved for the treaty's ratification on 16 September 1964, and the fate of the Arrow Lakes residents was sealed. This decision would cost the 2,300 people who lived along the Arrow Lakes their way of life, their well-being, and their long-accustomed environment.

The process leading to treaty ratification including the public hearings illustrates aspects of domicide discussed in chapters 3 and 4. The concept of benefit at local, regional, and national levels is revealed. For the victims, there is the sense of futility regarding any possibility of resistance; the pain of uncertainty; the difficulty of leaving land that has been worked for years to reach a productive state and the sense that nothing will replace it; the need for a quick settlement and compensation to allow those who were affected to find a new home, new land, and start the work necessary to get into production again; the victims' superior familiarity with the land compared with that of the officials; the official paternalism in negotiating the domicidal process ("we are going to take care of you"); the assumption that victims will always bend to the common good; and the giving of false promises. Given these feelings, it is important to examine the process of project development.

PROJECT DEVELOPMENT

In response to the impending flooding of the South Country behind the Libby Dam, the British Columbia government was held responsible for the procurement and preparation of land. To this end, the BC Water Resources Service issued a letter and brochure to residents of the area in the summer of 1968 introducing the program of reservoir operation, but unfortunately its delivery was delayed by a mail strike. Next, a preliminary *Guide to Property Owners*, which explained the benefits of the project and its extent, was delivered. The brochure indicated that the lake to be formed would extend sixty-seven kilometres into British Columbia, covering 7,223 hectares of land. Filling of the reservoir was to begin in 1972.

Many different agencies of the British Columbia government were to be involved: the British Columbia Water Resources Service as responsible coordinating agency; the Department of Highways for land acquisition and road relocation; and the British Columbia Forest Service for directing the clearing of the reservoir area. The BC Water Resources Service, Power and Special Projects Section, an engineering department, also became an unofficial ombudsman to represent people's interests (albeit from the capital city of Victoria).

Two small communities, Waldo and Newgate, were seriously affected, while other areas suffered adverse effects from the dam construction. The agencies directed an operation that resulted in the flooding of one hundred parcels of land and partially affected a further 120 parcels. There were 134 owners of the land parcels and twenty-five ranchers – about 200 people in all – who were to be affected as a result of the Libby Dam construction. The fate of these residents was discussed with them at a public meeting attended by 150 people in the school hall at Baynes Lake on 11 September 1968. Seven staff from the Water Resources and Forest Service, as well as representatives from the Highways Department, were in attendance (BC Water Resources Service 19 September 1968). Many assurances were given.

The meeting was opened by an official of the BC Water Resources Service with a description of the *Columbia River Treaty*, the policy on reservoir preparation, and a general timetable that contemplated land appraisals starting in September 1968, acquisition beginning in October 1968, clearing commencing in spring 1969, program completion by December 1971, and reservoir storage to begin mid-1972. The speaker stressed that "all property owners will be treated honestly and fairly and on the same basis" (*Fernie Free Press* 1968). This was followed by another address from the Water Resources Service in which the dam was described as well as the amount of land to be acquired. A map reserve, which ensured that crown lands would be retained, was to be placed around the reservoir to curtail speculation and to make certain that displaced persons could remain in the valley.

The Department of Highways property negotiator indicated that all appraisals would be undertaken by an independent appraiser from the American or Canadian Institute of Appraisers. Appraisals were to be undertaken, and then negotiations would occur when all the information was available. Full payment was to be made once the survey was complete; in the meantime, partial payment was available. Voluntary resettlement was expected, with compensation adjustments to be made later.

Following these presentations, there were only twenty-seven questions, since those who attended the Baynes Lake meeting had been advised that questions relating to private land holdings could not be answered at a public meeting. Many sought general information relating to the use or disposal of agricultural land and stocks, the flood level, and the fate of the community of Waldo; others commented on the delays in starting the preparation program. Only the first question and answer are noted as of particular interest to this discussion (Borneman 1993):

Question: Since we are forced off our land, is there any payment or compensation for the inconvenience created? How can a ranch be relocated in three months? We are caught in a trap for 20 years ... since 1966, we cannot ranch from year to year.
Answer: To be forced off the land is not in my jurisdiction. Three months to relocate is not correct. First amount paid is substantial, approximately 80%. It is recognized that land owners are being displaced. This is taken into account in the settlement.

Of significance in this interchange are a number of issues: the request for adequate compensation including compensation for disruption; the sense of entrapment caused by the lengthy time between project announcement and implementation; and the difficulties caused by the number of players involved from the government side in major projects. All of these difficulties contrast with the assurances of Senator Stanley and General McNaughton given in 1951.

Department of Highways officials were to visit the affected area in the fall of 1968 to begin negotiations for the acquisition of land. Even at this time, some properties had been offered for sale voluntarily. However, for others, the following words in a brochure were intended to give comfort:

Despite the fact that our Province will become richer, more stable, and more accessible as a result of the Columbia River Treaty projects, of which the Libby project is the last to be developed, we realize that the owners of the property affected by the Libby Dam Reservoir are deeply concerned.

For you, it will mean dislocation and change. Some of you may be pleased to sell your properties to us because it suits your plans to move elsewhere. Some of you, understandably, would prefer to stay where you are and have no desire to sell.

We cannot avoid the impact the development will have on your lives, but it is our intention to assist you as far as we reasonably can to ensure that the changes will take place as smoothly as possible (BC Water Resources Service Draft 1968).

The brochure went on to describe, in general terms, the land that would be needed, how the appraisal of land values would be made, how offers would be made, how expropriation would be used only when necessary, when it would be time to move, and how new communities could be established.

In contrast, the Arrow Lakes Project was much larger in scale and was subject to a more elaborate planning process. Construction of the project necessitated the acquisition of 4,376 parcels of land, of which about a quarter were owned by the Crown. The remainder were acquired from 1,350 private owners. Fourteen lakefront communities disappeared; 2,000 people living in the valley, 615 households, and 269 farmsteads and small ranches were affected. In addition, the land acquisition program affected many businesses and community institutions.

Remembering that there was very little onus on the project developer to care for the needs of those the project would displace (i.e., the water licence only required that the licensee review with the Comptroller of Water Rights prior to expropriation any matter where the licensee is unable to reach agreement with the owner or owners of land), the goals and activities of BC Hydro were significant. A redevelopment committee was established, with the goal to ensure the intention of the BC government "that any adjustments required by the Columbia developments shall be made in a fair and equitable manner" (BC Hydro 1961, 4). This five-person committee made contact with organizations and persons in the affected area and invited written submissions. In addition, their chair, H.D.C. Hunter, travelled from Castlegar to Revelstoke, explaining the proposed project to affected persons. At this time, BC Hydro also undertook a comprehensive census, title search of affected lands, and an economic study.

Once the treaty was ratified in 1964, BC Hydro published a booklet for property owners in which assurances were given that the Columbia Region would become richer, more stable, and more accessible and that the commission would assist residents "as far as we can to ensure that the changes will take place as smoothly as possible" (BC Hydro 1964, 2). In 1965, in *The New Outlook*, BC Hydro (1965) told the residents of the Arrow Lakes valley what their options were in terms of resettlement, including resettlement to a larger community (such as Nakusp, Revelstoke, or Castlegar), out of the valley, or to a new community. Three new communities were eventually created at Edgewood, Burton, and Fauquier.

Besides the activities described above, other actions were taken (BC Hydro 1966). These included: the publication of a *Columbia News Letter*, which kept the valley residents updated on the project; the publication of a regional plan and public discussions of this plan; the appointment of an Ombudsman, Chief Justice Colquhoun of the Supreme Court of British Columbia, who acted as a special Commissioner to advise those who believed their rights or interests were adversely affected; and the appointment of a former BC deputy minister of agriculture to help the aged and infirm to relocate; helping waterfront property owners to buy Crown-owned waterfront property elsewhere; holding community meetings; giving community planning and other financial assistance to schools and hospitals; arranging for appropriate donation of historic articles; performing economic and archaeological studies; providing significant community infrastructure in the form of tourist facilities, parks and beaches, marine facilities, housing subdivisions, new cemeteries, waterfront protection, water supply systems, roads, and power lines.

However, despite these efforts, as Premier Bennett opened the Arrow Lakes Dam in 1969 and renamed it after Hugh Keenleyside of BC Hydro, the hecklers "protested against honouring the man who had come to symbolize for them the tyranny of BC Hydro" (Hodgson 1976, 204). MP Bert Herridge, who fought for the preservation of the Arrow Lakes for twenty years, placed a quarter page advertisement in the *Nelson Valley News* on behalf of the Mourners of the Arrow Lakes Non-Partisan Association. The effect on people who were to lose their homes is still of significance.

THE PROCESS OF DOMICIDE
AND WHAT WAS LOST

While an assessment of what losing home would mean in the Arrow Lakes area was derived mainly from the water licence hearings, the story of some of those who lost their homes in the construction of the Libby Dam has been recreated from files and press reports. In both cases, comments heard in the course of the Kootenay Symposium meetings held at Cranbrook and Castlegar in June 1993, and described in a following section, added additional information. During the meeting in Cranbrook, particularly, in one of the smaller sessions, people shared their feelings about the past. For this observer (Smith), it was remarkable how they could speak so easily, and apparently without malice, of a past about which they clearly must hold some rancour. The notes taken can never adequately capture the faces of two of the women who told the story with a smiling calm.

To a large degree, it is a story of the enormous difficulties encountered by these people. Together, these sources tell a little of people's feelings about their homes, but more about the way in which they were treated. One caution is necessary here. As the files and newspaper accounts are reviewed, certain "stories" repeat themselves, a situation not unlike the traditions created within societies that depend on oral history. To some degree, this "perception" may be understood as "reality," for in such stressful circumstances, these stories become the only reality for the victims.

From the information gathered about the creation of the Libby Reservoir, one effect is more clearly demonstrated than all others: namely, the problems associated with the process. These problems may be generally described as the difficulty in finding a new home and appropriate land at an affordable price, problems associated with the method and length of negotiations, and the uncertainties surrounding expropriation and compensation. Negotiations were still ongoing towards the end of 1972, even though water flooding behind the dam was to start on 17 April 1973. In late 1972, there were still fifteen settlements to be reached, including four families whose homes would be inundated. It must be acknowledged that British Columbia's expropriation laws were less than adequate at this time and have since been improved; however, much of the problem can be attributed to the split in responsibilities for reservoir clearance between several government departments. The confusion that this engendered is summarized by Leo Nimsick, then minister of Mines and Petroleum Resources, when he wrote to the lawyer of a constituent: "I find many of these problems very difficult to analyze because there is so much hearsay evidence and very little written down in black and white. A big mistake was made at the start of this whole affair when the negotiations were not placed under one negotiator and with the power to buy the people outright by written agreement. Now I find that many of the people are saying they were promised things and the people to whom they refer deny any of it" (BC Water Resources Service, 1972–1981). For persons who were to be displaced, it was difficult to find a new home and to deal with disposal of existing assets given the conditions of uncertainty surrounding the Libby Reservoir project.

The length of time that it took for negotiations is very apparent. Presumably, some of this delay arose when residents chose to hold out, hoping for the highest price. Nevertheless, it is a part of the process of losing home that cannot be ignored. One person complained about how appraisal and settlement offers were handled, including the presentation of one expropriation offer during a wedding reception, and two offers by a lawyer who arrived late in the evening, apparently intoxicated. This circumstance led to an article in the *Vancouver Sun* titled "Rancher vows he'll stay put until dam floods his land," which stated: "The waters of the Libby Dam pondage are steadily inching toward the home of an 86-year-old pioneer rancher near here. But Henry Sharpe is determined to stay until he gets a satisfactory settlement for his flooded land. He is one of 13 individuals and three companies still trying to reach agreement with the government over compensation for their property ... 'My husband's going to stay put in our ranch house until the water comes,' said Mrs. Sharpe Saturday. 'If he

doesn't stay there the government might come in and burn it down –
they've done that to other people'" (Farrow 1972).

Underlying all of the difficulties for those persons who did not settle until the very last moment was their firm belief that the government did not appreciate the value of their land. Appraisals were based
on specific criteria (BC Water Resources Service 1972–1981). While
the appraisal guidelines may have tried to create a fair process, many
of the ranchers eventually hired an independent appraiser because the
government-appointed appraisers did not appear or their valuations
were considered unfair: "I have had no attention paid to my appraisal,
instead I have had to deal with people that appear to have no idea of
the value of ranch land, refuse to meet with my appraiser, and prefer
intimidation type of tactics when dealing with the people" (BC Water
Resources Service 1969). Further: "And the people who haven't settled can't afford to for the offers they've received. There are discrep-
ancies of $30,000 to $40,000 between the prices offered and what the
people believe their property is worth" (Farrow 1971). But to counter
this, the lawyer for the Highways Department, who grew up in the
valley and recognized the bitterness of the "victims," insisted that the
prices were fair: "The trouble is that people have become bitter over
being uprooted and not knowing where to go. But simply because
someone is sentimentally attached to a place doesn't mean that we can
give him an extra $50,000 of taxpayer's money ... Sure, I'll try my best
to work out some agreement with a landowner. But it's got to be justified. You can't put a price on sentimentality. Its simply a case of public good taking priority over individual rights" (Farrow 1971). This
statement seems to underlie the apparent belief by many that, while it
is difficult to give up your home, such an action is acceptable, given
adequate compensation. In the end, given the momentum that devel-
ops around major project construction, it is often this question of
compensation that forms the basis for resistance by the victims rather
than any feelings about home.

Resistance in these circumstances is frequently led by one or two
individuals. A speaker at the Kootenay Symposium meetings, held in
Cranbrook on 4 June 1993, paid tribute to Lloyd Sharpe, describing
his "valiant effort" on behalf of the ranchers, but, in the words of the
speaker, "it was bigger than all of us." Sharpe was spokesman for the
ranchers and fought against the project for many years. In particular,
he objected to the length of time that the negotiations for expropriation took and the values placed on the ranches. When asked to discuss
the benefits of what had happened, one delegate at the Kootenay Symposium said that it "brought out how little power people have" and
"the need to tell future generations to fight for everything ... to put in

a wide open process for negotiation ... and legislation which will protect individuals – the common thing is for the common good." The speaker felt that the "common good" would always prevail over the rights of individuals.

In general, the loss of home to those affected by the Libby Dam meant loss of a beautiful place, livelihood, family, identity, familiar surroundings, and community. It also had a negative effect on their health. To put it most simply, "it was a beautiful valley" (Kootenay Symposium 1993 *c*). One reporter suggested that the 300-acre Island Ranch, owned by Lloyd Sharpe and founded by his father in 1917, was the most beautiful place of all, with its "lush meadows and woods abound[ing] with deer, moose, elk, and bear" (McCandlish 1971). Another spoke of a property where, "from the living-room window, you look across a little valley and see cattle on their winter grounds. A creek runs through the valley, which is dotted with a few small lakes" (Jacobs 1968, 58).

Given the rural nature of this area, it is not surprising to learn that loss of livelihood was a major concern, particularly for the twenty-six ranchers of the area who became "victims of the great Columbia River Treaty" (Jacobs 1968, 10). This situation must have been particularly difficult for the men who had come home from the Second World War to settle in the valley, only to see the Libby pondage occur (Kootenay Symposium 1993 *c*). In 1951, United States Senator Stanley had told the ranchers: "You will not lose a dime – we will pay you big American dollars" (Jacobs 1968, 3). But the ranchers claimed that they had been losing for a long time prior to the dam construction. Because the project was under consideration, the government recognized that expansion of the ranches through crown land would cost money when it came time for expropriation. Hence crown land sales were terminated. In reaction, Jack Aye, the community leader who had ranched in the Sand Creek area on the borders of the Kootenay River for thirty-one years, and whose comments at the International Joint Commission hearings seventeen years earlier have been quoted at the beginning of this chapter, stressed the cost of uncertainty:

After World War II, up till quite recently, ranchers anywhere but here could expand. Some of them bought crown land at $3 an acre, which they could develop; others got additional grazing permits, or leases, or S.U.P.'s. We couldn't do anything. If we had been smart, we would have sold out then and taken our chances somewhere else. We kept thinking next year we'd know what the score was and we might be able to make a deal.

Another thing works against us ... this area is dead if you try to sell it. Who would pay any real money for a place along here, knowing that it could be flooded at any time? Who would lend you money to buy a place? The few places which have sold, have traded under value – really at ridiculous prices. When the government does come to expropriate, they're going to look at these low prices and figure that's what our land is worth. And how can we put a value on what is lost in the last 20 years because the government hasn't let us expand our units? (Jacobs 1968, 54).

Besides loss of the physical structure, loss of home also meant loss of family and community. Persons attending the Kootenay Symposium recognized that before the reservoir was formed, there was a community that was of importance to its residents. One person stressed the special generosity of the people of this area. Another felt that breaking up the social fabric of the community should have a price tag, too, "just as it did with the restoration of communities in Los Angeles after the race riots." But many people did leave the area entirely and now are forgotten. A few people remain, but they feel that new people moving into the area do not become so involved in the community. One speaker, from the generation after those directly affected, noted that many of their relatives lived south of them, and when the valley was cleared, they were the only ones left. Family histories were lost (Kootenay Symposium 1993 c).

Loss of home is particularly hard for those who have lived in the same area for much of their lives. One couple affected by the reservoir had lived in their ranch house for fifty years. Another said: "All my working life, I have lived right on this ranch. Where are you going to find a place like the one you grew up on?" (Jacobs 1968, 54). Others lost only half their land, but it was difficult to begin building a new home on the same lot on which they had lived all their lives. One person felt that "It looks as if the government's policy is to drown us out like gophers and then if any survive, keep making excuses until we die of old age" (BC Water Resources Service 1972–1981). It was perhaps easier to just give up in these circumstances: "Sure, its hard uprooting yourself after living in the same place for years," he said, "but I don't have much longer to go, and I'll let the younger men fight the government" (McCandlish 1971).

For older persons, the effect of stress and uncertainty was particularly evident: "There have been some elderly people around here who died of worry," she said. "Ownership seems to mean nothing to the government. It's so unsettling not knowing what we are going to do in the future" (Farrow 1972). And "His health is failing and he is worrying himself sick over our situation. He repeats 'if I should die tomorrow, you haven't even got a home,' which I know a lot of people don't have. But we did own a comfortable home ... and did not owe a cent to any one. We have had this terrible debt over our heads and heavy on our hearts for 1 1/2 yrs. now, and it is surely showing" (BC Water Resources Service 1972–1981).

Some people became ill or were very disturbed by the negotiation process, finding difficulty in coming to terms with the reality of the situation (Kootenay Symposium 1993 *c*). A Forest Service crew member, who was sent in to burn the evacuated houses, tells of one occupant who simply left all his furniture and departed the night before his house was to be destroyed, compensated either inadequately or so well that his future relationship with his neighbours might be called into question (Bugslag, 1993). A negotiator acting on behalf of the ranchers said: "It is a deplorable situation. They are driving the hearts and minds out of those poor ranchers" (Farrow 1971). At the Kootenay Symposium, it was suggested that compensation should be given for the stress of forty years that had hung over their heads and those of their children (Kootenay Sympsoium 1993 *c*). In the final analysis, however, another delegate recognized that money doesn't replace the things you hold dear (Kootenay Symposium 1993 *d*).

Review of the process and meaning of domicide in the Arrow Lakes situation reveals many similar themes to those expressed in the Lake Kookanusa area affected by the Libby Dam. Because most of the information was gathered from the hearings' transcripts, the perspective provided is frequently from people who still may have felt that they had some hope of changing their circumstances. Three themes are identified: namely, the plague of uncertainty, the feeling of injustice, and finally, the sense of bereavement and bitterness felt. Over and over again, people at the hearings related feelings of uncertainty, feelings which affect both individuals who seek certainty as to their future and the community where a sort of "planning blight" sets in: "I understood when a man came in there first and spoke about the dam that this would take place, and in the very near future, but it has dragged on and dragged on until I don't know what to do any more, whether to do any more improvements or not, although I have done some things, but things are pretty much at a standstill" (Comptroller 1961*c*, 833). And "Every place I go, I make my living up in the Arrow Lakes for over 25 years, and through the indecision of our Governments, I have been unable to get any work. Nobody wants to repair anything. Every place you go, the people are really nervous, and only people who make something out are doctors and chiropractors" (Comptroller 1961*c*, 820). Finally, people could only hope that "what you have to do, and feel is your duty to do, may it be done quickly and put an end to this uncertainty, this pall which has descended on this area where no decisions can be made" (Comptroller 1961*b*, 642).

The prospect of domicide might have been better accepted if there had been some certainty around the issue of compensation. Certainly Wilson believes that the whole Arrow Lakes program was overshadowed by the issue of compensation (Wilson 146). Hugh Keenleyside, speaking on behalf of BC Hydro at the water licence hearings, had made it clear that no decision had been made about a fixed rate of payment and stressed that each case deserved individual compensation. He suggested that each family would be paid the market value of their land plus a bonus of 10 per cent for forcible taking. But the problem was that the value of the land in the area had been depressed because of the years of uncertainty over project construction, and sometimes it was just not possible to find comparable agricultural or other situations. And if an owner held out, would that ultimately result in a higher settlement? (Waterfield 1970, 45–7; Wilson 31–2). Concerns of residents of the area also included the feeling that they would not be fairly compensated, the belief that "No amount of money would ever repay us for what we would lose, our home, our livelihood, and our whole way of life," or that it was impossible to compensate for future value (Comptroller 1961b, 652, 644; Comptroller 1961c, 800).

During the water licence hearings, people praised the beauty and peace of the Arrow Valley, which provided the setting for their home:

The dam will completely destroy practically all existing and potential amenities of this 150 mile long valley, whose elevation, soil, climate and beauty are second to none ... but [for] visitors and tourists who love the beauty and peace of our Arrow Lakes, the shores and beaches would be lost for hundreds of years (Comptroller 1961a, 462).

We have been looking forward to move to that place and have a nice quiet life after 37 years working on the smelter and all kinds of noises ... but everything is kind of shattered now, and that is actually my only complaint (Comptroller 1961c, 831).

I never seen anything like [this valley on] three continents, in Europe, in Australia, and in this continent. The situation is absolutely ideal. It is warm, there are beautiful mountains, there is lovely forest, there is good land, the whole area is well protected against gusts of wind (Comptroller 1961a, 446).

The sense of violation was stated most dramatically by Mrs. Alma Jordan who said simply: "Mr. Comptroller, ladies, and gentlemen, I protest the mutilation of our beautiful valley" (Comptroller 1961*a*, 409).

However, the valley was more than just a beautiful place as a setting for home, for, with its small settlements and natural resources, it provided a living and a source of community life. With the threat of the Columbia development, the area stood still from the 1950s onward. Because of this, many of the people in the area were older, those "who had lived in the valley for decades and for whom the whole landscape, natural and human, was a comfortable tapestry of things familiar" (Wilson 1973, 10). It was the "humanity of the landscape, evolved over so many years of settlement" that disappeared with the coming of the dam (129).

It is also this humanity of the landscape that brings us to a review of what home meant to the residents of the Arrow Lakes. The best source for such information is the thousands of pages of transcript from the water licence hearings, even though such evidence was frequently not admissible in the eyes of the comptroller of Water Rights, described as an "omniscient presence" by one of the people of the valley (Comptroller 1961*b*, 655). From these transcripts, published sources previously described and statements made at a recent "revisiting" of these events (Kootenay Symposium, 1993 a,b,d), five main themes appear relating to loss of home, and three relate to feelings associated with the process leading up to their loss of home as well as to their sense of loss.

The themes relating to loss of home were: 1) similarities to fighting for home/nation; 2) loss of environment; 3) loss of entity/community/place /final home; 4) loss of land/security/property rights; and 5) loss of initiative/health/effect in old age and emotional disturbance or loss of heart. Feelings about the process of loss of home included: 1) uncertainty regarding expropriation and compensation; 2) sense of injustice in relation to those who benefit; and 3) a sense of bereavement and of bitterness.

References to home at the water licence hearings were often coupled with references to nation. Since two world wars were very much part of people's memories at this time, the comparisons between fighting for one's country and for one's home were easily made, particularly for those who fought in the war, came back, acquired land under the *Veteran's Land Act*, worked that land, and now found that they were to be removed from it: "If this High Arrow goes in, my place will be completely flooded. I will have no land or anything left, no home, and I have lived on the place now for fifty years. I was practically raised in the Arrow Lakes. I was only four when I came here. Furthermore, I joined the Army to fight for this country, and I believe that every returning soldier should have a fair chance, not to be driven from his home" (Comptroller 1961b, 741). The spokesperson for the Nakusp Women's Institute stressed the relationship between home and nation, citing similar feelings of sentiment toward each, as well as recognizing the stabilizing influence that each provides: "The 25 to 30 members of the Nakusp Women's Institute ... do have and are concerned with their motto, 'For Home and Country.' It seems now that these two are closely bound up in each. Love of home and country, and the desperate measures taken to protect them, has sometimes been called sentiment. If it is, then it is something very strong ... In our opinion, the very words 'home, country, sentiment, patriotism,' call it what you will, is still one of the most stabilizing influences we have left today. It provides a counter-balance to the changes and uncertainty the world over" (Comptroller 1961a, 461).

The most powerful of the presentations was that of the Reverend Pellegrin, deacon of the Anglican parish in the Nakusp area, who stressed the importance of home in people's lives and their identification with home more than nation and, in so doing, emphasized the magnitude of the expected loss: "In times of national stress, we are urged to show in visible ways what our fatherland means to us, to pay the price for self-determination and, if necessary, to die for it. You can't expect us now to give away what many have died for just because the government feels that it is a convenient way of making some $65,000,000 or thereabouts. To each of us, our home is the most important place on earth, and it is this which we seek to preserve, not some nebulous thing called a nation" (Comptroller 1961a, 509).

Just as strongly felt was loss of place, community, and identity. At the Nakusp meeting of the Kootenay Symposium held on 2 June 1993, these communities were remembered. When asked to define the past impact caused by the construction of the reservoir behind the Keenleyside Dam, people listed loss of small community lifestyle, friends, and roots. The Story of Renata 1887–1965 tells of that community life that was lost: church services, the Women's Institute "Dollar Teas" for the upkeep of the cemetery, and the "bees" to butcher the pigs in the fall – each family attending ten or more bees within a few weeks (Warkentin and Rohn 1965).

Loss of community also meant loss of the final home for the ancestors of the flooding victims. The fate of cemeteries became an issue at the hearings when it was suggested that they might be covered by concrete. As one person said: "It is a disgrace to disregard the work of the pioneers of this valley, those who have passed on, to cover their graves with dirt, driftwood, and filth of all kinds every year, when there is no sensible reason to do so, only for money" (Comptroller 1961*a* , 415). In describing the detailed steps taken to close cemeteries, Wilson acknowledges what an emotional subject this was. He remembers one young woman who was able "to stir a crowd by crying, 'We are not just to be thrown out of our homes. Now it appears we can't even die in peace!'" (Wilson 1973, 90–1).

Loss of home and loss of "the concentrated efforts of our hearts and hands" (Comptroller 1961*b*, 644) was most difficult for those who had cleared the land with their own hands and stayed, often when there was no profit in so doing, believing always in the valley's future potential: "My late parents and myself took possession of this property in 1924 and have improved it with good buildings, cement walks, propane gas installations, rock wall, gardens, fruit trees and planted the street with acacia trees, now grown to enormous size and height. The street for nearly two blocks is lined with these trees, and also maples. It took many years to bring them to this beautiful and useful stage" (Comptroller 1961*a*, 489).

And what losing this meant – the loss of memories and of identity – is explained in the following:

After six years of hard work and after having made something out of nothing, it is not a very pleasurable position to see the thing ruined and go under water, even if you are paid for it. If you work, it is not only for doing something for yourself, but to do something for the country, and if it is ruined, it really hurts.

I have my personal experience what it means to lose what you have. I had a very nice property in Europe, in Poland, 3,000 acre property which we farmed for four generations. I had only two hours time to run away from it when the Russians came, in order to keep my head where it belonged. It is not the financial loss that hurts. What hurts is that you are losing the land on which you worked, where you know every bit of it, where you get accustomed to it, you know how to farm it. It is full of remembrances, of your failures and successes. It becomes a part of you (Comptroller 1961*a*, 446–7).

The harder edge of loss for those who have worked to build up a home and, in the case of the Arrow Lakes Valley, often their livelihood, is loss of security and property rights. A sense of helplessness is created: "So far, I have been working my head off to get far enough to have a roof over my head, or even my family, if necessary, but if they take it away from me, I have nothing" (Comptroller 1961*c*, 821). A sense of desperation is expressed:

I am now 61 years of age, and up until the last twenty years or so, have not been in a position to help myself financially to any extent owing to various conditions. During the last twenty years, I have put practically all my savings in earnings into my home, of which I am very proud. My wife and I are entirely contented where we are, and I have arranged our house and grounds to suit our wishes, and at the same time, stay within our means. Therefore, considering our age and contentment, we would not sell willingly to anyone. In other words, the only way we would move is to be compelled to. Apparently, that is what may happen. No one knows the meaning of this except those that are vitally affected. At our age, we do not feel like starting over again to get our home and grounds prepared in the same condition as they are at present (Clough 1961).

Loss of property rights is, of course, a serious issue. The spokesman for the Farmer's Institute stressed the importance of tenure: "WE OBJECT to this water licence on moral grounds. More than any other section of the community, farmers set great store by their indefeasible titles. Farming is a long term business based on security of tenure" (Comptroller 1961*a*, 548).

Loss of home may also signal a loss of heart, loss of initiative, and even loss of health: "No, a few hundred or a few thousand dollars for each family will not buy a real home, for the people's hearts will be lost in the flood. There can be no love if hearts are gone, just death in our valley" (Comptroller 1961*a*, 464; *b*, 638). And: "The small man was encouraged to play a large part in this development. It is the small man, such as my husband, who has been the backbone of this country, and I deplore seeing much of the initiative taken away from them." Another person spoke of the difficulties of salvaging materials from her old home and told how her husband took ill and died one year after they moved. Yet, still she hoped to make the best of her situation (Wilson 1973, 144). At the Kootenay Symposium, one speaker said: "I have been waiting for twenty-five years for this opportunity." He then told how his mother, who lived in one of the first communities to be flooded, lost a store, six lots and a home on a two-acre waterfront lot, all for compensation of $47; she subsequently had a nervous breakdown (Kootenay Symposium 1993*d*).

Waterfield (1970, 48-9) writes of an individual who, because of his role as a real estate and insurance agent, probably stood to gain while others lost, but who spoke strongly at the hearings of how the dam "would smother the spirit and initiative of many of our people." Waterfield concludes that the speaker was emotionally distraught but suggests that so were most people faced with eviction: "How could one be otherwise and claim membership in a sentient human race? While one cannot prove it, many people firmly believe that elderly persons who, later, through the procrastination of the servants of the BC Hydro and Power Authority, witnessed the destruction of their and their neighbours' homes, were emotionally disturbed – fatally."

Wilson (1973, 110–11; 144–5) suggests that older persons, of which there were an unusually large proportion in this area, were often the least able to cope with losing their homes. Not only were they not able to start over again building or working the land, but also they frequently lost the legitimate "expectation of living the rest of my years in familiar surroundings in a home that I had helped to build among groves and flower beds that we had built up gradually over the years. No amount of money can compensate me for this property, and I am now too old to start over again." For many, moving from their home would also mean loss of independence (Comptroller 1961*a*, 510; 1961*c*, 823, 838). Older people often dislike any sort of change, never mind the massive transformations that they had to face here.

In fairness, it must be noted that for some people, there may have been an advantage to moving. Based on a survey of eighty persons over sixty years of age who were to be displaced, Wilson found special problems including low earning power and therefore a requirement for a low cost of living, as well as physical disabilities, lack of mobility, and dependence on neighbours for chores and transportation. For these people, it is suggested that the move to the village of Nakusp or other larger communities may have been welcomed, particularly for the access to medical services. In these circumstances, with no follow-up survey of older persons possible, it can only be suggested that while living standards may have improved for some, the loss of long-time homes would be a terrible thing to endure. Wilson concludes that "forced removal from a familiar environment is an inescapably brutal act which should in no way be glossed over" (Wilson 1973, 145).

WHO BENEFITED FROM
PROJECT CONSTRUCTION

The question of who benefited by the construction of the Columbia Basin projects is an important issue. Persons affected by the Libby Dam project were particularly concerned about the need for fairness in the process, given that it was others who benefited. Many felt that: "We're being thoroughly and completely shafted by the government of British Columbia" (McCandlish 1971). Reflecting on this, thirty years later, one person noted that in the 1960s, bigger was always better, and it was this belief that set government priorities; now, our value system is changing (Kootenay Symposium 1993 c). But returning to the time when the dam had just been constructed, after hearing Donald S. Macdonald, Canadian minister of Energy, Mines, and Resources, address the dedication ceremony that day, one of those who had lost their home to the dam wrote: "Do you think it is justice to take away a man's livelihood and let him starve to death when you agree that the *Columbia River Treaty* was such a benefit to the rest of the people?" (BC Water Resources Service 27 August 1975)."

Another wrote to local MLA Leo Nimsick regarding a proposed "Access Act," demonstrating the sense of trespass, felt both when someone crosses your property illegally and when property is taken for dam construction: "But when there is a *reason* to take care of *a few* people with not enough votes to interest anybody, there isn't any money, so the story goes ... The act passed regarding Libby basin gives powers "to take" any property needful but not one word about payment – no money mentioned" (Nimsick Correspondence 2 April 1968). Nimsick's response represents a politician's answer to the question of who benefits. He replied that as a member of the opposition, he had only limited influence on government policy. He had done his best to explain the wishes of the people to the government, and:

contributed something, I hope, to a solution and which will assist the farmers during the period that is coming up in the near future when these properties will have to be bought out ... It is difficult to give you a complete picture of what our rural representatives must try to do. He has got to be interested in the common good of all the people, not on just one section, and while some people may think I have failed in this regard, it has not been due to want of trying (Nimsick Correspondence 4 April 1968).

Or, put another way, on behalf of an individual who, with his family, had ranched in the area for approximately fifteen years, on a farm that had existed in the area since 1912: "It does seem fair to suggest that when a man is expropriated innocently, as was the case, then everyone should move over a little to let the victim carry on as he desires" (BC Water Resources Service 8 January 1973). In the colourful words of one Arrow Lakes resident: "Too much luminous paint [was] being used on the benefits of High Arrow – too little emphasis on the detriments and the losses through destructive flooding"(Comptroller 1961b, 701).

There was another side to this story, however, perhaps best spoken by someone who might benefit. A real estate agent and local mayor in the area was quoted as saying: "It's a shame that good farming land is to be flooded. But I think the benefits of the dam will outweigh the negative factors" (McCandlish 1971). And even those affected directly can apparently view the situation in perspective. A speaker at the Kootenay Symposium remembered a rancher telling him that "a few have to suffer so many can enjoy a better quality of life" (Kootenay Symposium 1993 *c)*. While the British Columbia government saw financial advantage in the project construction and the federal government saw the opportunity to be a good neighbour, the residents of the Arrow Lakes saw only benefit to the United States.

Examined in isolation, it would appear that feelings engendered by the *Columbia River Treaty* projects might have been sufficient to at least sway the choice of location for projects in the Columbia Valley, particularly given that alternative choices were available. We need to understand why resistance failed in these situations and who did benefit from dam construction. In fact, the issue of benefit in the "public interest" or "common good" took on a whole new meaning in the context of the Columbia negotiations. Any resistance that did occur was simply not seen as significant against the background of the negotiations for the *Columbia River Treaty*.

The *Columbia River Treaty*, as it finally emerged, recognized that the Canadian treaty projects – Duncan, Keenleyside, and Mica – as well as the Libby Dam in the United States, would increase the energy output and dependable capacity of the American power plants (British Columbia *a* 1993, 5). In return, Canada was entitled to one-half of the additional power generated by the power plants as a result of storage in Canada. The calculation of this entitlement is determined six years in advance and is based on the amount of power that would be produced by the American plants with and without the Canadian storage. Because this power was not required in the 1960s, these benefits were sold back to American utilities for a thirty-year period for US$254 million. From the British Columbia government's perspective, Swainson (22) identifies three overall benefits that accrued to the BC government based on this decision: 1 fiscal restraint and the elimination of provincial debt, 2 the establishment of a reputation as a government that would get things done, and 3 projecting an image of large-scale thinking.

While the second and third of these benefits may simply relate to the creation of a good political image, that image was the only benefit the provincial government of the day got from the *Columbia River Treaty*. While there was financial advantage, it was more than eaten up by other new hydroelectric construction. It is only today that a fourth gain, from Bennett's prediction regarding the future increased demand for power, has the potential to be fully realized as the downstream benefits return to Canada (Swainson 1995).

But it is interesting to examine the issue of benefit at the time of dam construction beyond the creation of an image. *The Columbia River Treaty* projects certainly had a public interest component (social and economic gains as discussed above, including the more tangible one of flood control). However, there were also advantages to those who held political, bureaucratic, and individual power. Not unimportant in considering the latter category is the creation of a sense of achievement and therefore enhanced self-esteem for those in power. While a discussion of such an intangible effect is difficult to portray, it does occur. When individuals are successful in achieving control, just as in the business world when the rewards are more frequently monetary, there are benefits in terms of prestige and knowledge of goals achieved. This sense of self-esteem contrasts sharply with the losses experienced by the victims of domicide.

The above discussion has been based largely on documents that existed from the time of the treaty negotiation and project construction. Common threads that run through these documents were the significant delays and uncertainty that were experienced by the victims of the Columbia River projects. Much of the delay was caused by the lengthy period for study of these projects and for negotiation of the final agreement. Much of the negotiation was taken up with the issue of whether British Columbia would adequately capture the financial value of the "downstream benefits" such as protection from flooding and access to electrical power that accrued to the United States as a result of project construction. Nine per cent of British Columbia's entitlement was expected to return in 1998, 46 per cent in 1999, and 45 per cent in 2003, based on the date of completion of the various Canadian dams. The entitlement is now expected to be much larger than was predicted in 1964, and the Province of British Columbia is considering alternatives to receive their entitlement recognized in the treaty. These alternatives include: agreeing to a return of power at sites other than Oliver (as originally agreed), permitting the United States to pay for alternative power sites in British Columbia, or reselling some or all of the power (British Columbia 1993b, 3,6).The issue of the return of the "downstream benefits" to British Columbia leads directly to a consideration of initiatives begun by the people of the Kootenays and the provincial government in 1992. In particular, these initiatives are a manifestation of Smith's reminder quoted in chapter 6: "When places die ... they are often believed to be no longer worthy of attention. But such places are still of importance to their residents, and therefore when 'death' is a consequence of public policy ... the implications have to be addressed as a matter of social responsibility" (Smith, 1992).

PRESENT AND FUTURE

The existence of downstream benefits from the Columbia River projects to be returned to British Columbia and the belief of the people of the Kootenays that these benefits should finally return to the area that suffered most was the impetus for the Kootenay Symposium in Castlegar, BC in June 1993. In order to provide a unified voice for the people of the Kootenay region, the Columbia River Treaty Committee was formed in 1992 to represent "the 260,000 people within municipalities, Regional Districts and Native Councils in the Columbia River Basin who have been affected by the building of hydroelectric and storage projects on the Columbia River and its tributaries as allowed by the *Columbia River Treaty*" (Columbia River Treaty Committee 1993). The Committee included two members of each of the five regional districts (local government) and two members of the Ktunaxa-Kinbasket Tribal Council. The goals of the committee were to forge a reasonable partnership with the province in the negotiation process, a partnership that is mutually beneficial, and to allow regional representatives to have a direct voice in negotiations and be active participants in the decision-making process.

At early meetings held in smaller centres throughout the region, there was specific recognition that there was loss of home and community as well as hardship caused by litigation to settle the claims, but there was little public call for restitution to the victims of domicide. A much larger gathering, the Kootenay Symposium, was sponsored by BC Hydro, the provincial government, and the Columbia River Treaty Committee in Castlegar from 18 to 20 June 1993 to discuss the future of the Columbia-Kootenay Region and to plan for the use of the downstream benefits. This symposium was attended by three provincial government ministers, the Columbia River Treaty Committee, representatives from BC Hydro, and other provincial government departments. Also in attendance were nearly one hundred delegates hand-picked by the treaty committee and the provincial government to represent specific viewpoints. The symposium began with an open house to familiarize meeting participants with the treaty, downstream benefits, the operations review currently under way by BC Hydro and an economic overview of the region. In the evening, a retrospective presentation was made by James Wilson, author of *People in the Way,* and delegates to the symposium who focused on the benefits as well as the negative impacts of dam construction. This was the major opportunity to relive the past, and the remaining two days of the symposium were used to look to the future of the region.

As it addresses questions of concern to this discussion, the Columbia-Kootenay summary report from the symposium dated August 1993 brings forward a number of specific proposals and includes general assurances from the provincial government that it will pursue the recommendations of the symposium (Salasan Associates 1993). Two statements are of particular interest here. The first, under the heading Need for Redress, stated that: "focusing too much on redress can be destructive and cause division." The second is the government response that "both government and BC Hydro must admit past mistakes, redress grievances, and take actions to involve people in future decisions" and that "the trust may be lost, but trust of future generations can only be built on actions, e.g., allocation of all or a portion of downstream benefits to a regional management authority" (Salasan Associates 1993). These statements suggest that there is a recognition of the need for redress, but when viewed in the context of the whole symposium report, this redress is seen to be interpreted as various forms of regional benefit.

In summarizing the options for immediate use of the downstream benefits, the interim summary of the conference proceedings found six possibilities: 1 personal compensation, 2 social development, 3 environmental mitigation, 4 investment analysis, 5 sustainable planning, and 6 authority development (Cornerstone Planning Consultants 1994, 1a). On 8 March 1995 the return of a portion of the downstream benefits ($1 billion) to the Kootenays was announced. On 16 May 1995 these hopes were dashed as the United States government suspended negotiations on the non-binding agreement that they had negotiated nine months earlier because power valued at $45 a kilowatt six months previously was now worth $20 a kilowatt. The British Columbia government nevertheless announced that they would honour their commitment to a return of part of the downstream benefits to the people of the Columbia Valley.

The Columbia Basin Trust was formed, and in July 1997, a long-term management plan was completed, which included an objective that the trust should "advocate for unresolved compensation claims" (Columbia Basin Trust 1998a, 12). In June 1998, a news release (Columbia Basin Trust, 1998b, 1) indicated that the trust had established a program to assist original landowners who were affected by dam construction under the *Columbia River Treaty* to make their case for unresolved compensation claims to the BC ombudsman. Eligible claimants were to be provided with up to $2,500 each toward the reimbursement of expenses incurred in making and substantiating a claim to the ombudsman.

LESSONS LEARNED

This chapter has discussed loss of home through the creation of the hydroelectric storage reservoirs in the Arrow Lakes Region and the South Country of the Kootenays. The primary purpose of the chapter was to test the validity of the concept of domicide through an empirical study, and, in particular, to examine the following: the meaning of home and loss of home; the parameters of domicide including process, motive, and benefit; the reaction to loss of home; and the means by which those affected were planned for.

Reflecting on all the material examined, there is clearly a significant degree of similarity between the findings of previous chapters and those of this empirical chapter. It is clear that home for the residents of the Arrow Lakes and the South Country meant choosing to live in a beautiful place with a strong sense of community and for those who worked the land, a strong sense of attachment and identity. Loss of home meant loss of environment, identity, community, livelihood, a beautiful place, familiar surroundings, final home (grave), land, security, initiative, and health. The loss of historic continuity, the sense of being rooted, was also felt. Both valleys had been settled for over fifty years, and while they might have seen better economic times, there was a quiet sense of prosperity to come.

The process of domicide in the Columbia River Basin has been compared to the conclusions of chapters 3 and 4. Many aspects are similar: project construction justified in the public interest (socio-economic benefit and flood control); disempowered victims (years of uncertainty, long-term grieving, and some effect on health); paternalistic bureaucracy (particularly in the International Joint Commission hearings); a sense of injustice on the part of the victims in relation to those who benefit; a sense of bereavement and bitterness; and problems with appraisal and compensation.

Particular insight has been provided about loss of property rights. This was very serious for those who lost both home and livelihood. As Million (1992) demonstrated, there is no more significant tie to home than that which exists for those who work the land. Finally, it is again found here that, once slated for destruction, home became primarily a marketable commodity to be given up in return for what people hope will be fair compensation. Perhaps it is in these moments that the victims of domicide, unconsciously, attempt to achieve the true monetary value of their "embodiment of identity" (Million 1992, 153). In the circumstances of the case studies, it is apparent that for some, this value was not, and perhaps could not ever be, realized.

There were also some specific differences from the discussion developed in earlier chapters. In particular, in the Arrow Lakes, the loss of an environment of great beauty was frequently cited and is mourned even today. The sense of home as centre and the effect of loss of home on identity were not as clearly established. However, aspects defining centre were evident: there were references to refuge, freedom, possession, shelter, and security. As to identity, from persons who spoke at the Kootenay Symposium, their experiences over thirty years appear to have caused them to forge a new identity, both for themselves and for their children.

The differences between these case studies and the findings of previous chapters may well be explained by the following factors:

- This research was almost totally based on the examination of written records rather than interviews, such as those undertaken by Million (1992), and therefore, the opportunity to prompt subjects to provide more details was lacking. As a result, less depth is found in discussions of meaning of home.
- Most of the records reviewed were from a time prior to or during the final loss of home, a time when naturally there would be little opportunity for deep reflection on what loss of home meant.
- The records in existence are from a very small sample of people who actually lost their homes, and usually from persons who resisted the destruction of their homes.
- In many cases loss of home meant loss of both home and livelihood for farmers or ranchers, rather than simply loss of a dwelling. The emphasis is, therefore, often on loss of land rather than the physical structure of home.

It is noted, however, that comments made at the Kootenay Symposium by those who lost their homes thirty years earlier echoed the thoughts of those who spoke at public hearings and through letters to the Water Resources Service at the time of project construction.

In reviewing the process by which people lost their homes, there were also some factors that were more clearly delineated than in the previous chapters. These included:

- public hearings that did not really deal with the people in the area (IJC hearings, 1951) or were undertaken after the real decisions about project location were taken (Comptroller of Water Rights, 1961) and therefore enhanced the sense of grievance and bitterness felt by those who lost their homes

- the longer period of uncertainty prior to project construction which affected individuals who wanted greater certainty as to their future ("You are worse scared than hurt") as well as lowering property values when property deteriorated while treaty negotiations were under way
- the sense of entrapment given that period of uncertainty
- the sense that people were always waiting for a message from afar (Victoria, Ottawa, Washington, DC), which would seal their fate
- the difficulties that arose in the South Country, given the number of players involved in reservoir clearance (various ministries, many private appraisers)
- the apparently constant belief that the victims must bend to the common good – hence the lack of focus on the victims, particularly in the South Country.

In regard to the latter, this chapter has explored, in more detail than in chapters 3 and 4, the issue of who benefits from project construction. Certainly, there was benefit to the communities and rural areas that no longer experienced flooding, to American industries that required electrical power, and to the British Columbia government and therefore to the Kootenays in the future, in the form of the return of the downstream benefits. It is also suggested that gains included an increase in the self-esteem, based on the achievement of goals, of a number of people holding political, bureaucratic, and community power – a concept not dissimilar to that of the "evangelistic bureaucrat." Finally, this chapter touches on the planning efforts made by BC Hydro to provide mitigation to those who would lose their homes: meetings, advocates, publications, surveys, community relocations, community financial assistance, and new infrastructure.

In this chapter, we have attempted to demonstrate empirically the salient characteristics of the process of domicide. To a lesser degree, we have demonstrated what loss of home means. Contact, which was made with persons during the Kootenay Symposium in 1993 and 1994, confirms the underlying sense of bitterness that lingers in this area even after thirty years, bitterness that is tempered by the desire and the strength for new beginnings and the recognition that the public interest can be viewed as a positive element. Nevertheless, it is clear that more could have been done for the "victims" of the Columbia River projects. In this connection, chapter 6 will examine both the process of planned change and the means by which both planners and those faced with domicide can better deal with proposals for major projects.

CHAPTER SIX

The Nature of Domicide

It is not what they built. It is what they knocked down.
It is not the houses. It is the spaces between the houses.
It is not the streets that exist.
It is the streets that no longer exist.

It is not your memories which haunt you.
It is not what you have written down.
It is what you have forgotten, what you must forget.
What you must go on forgetting all your life.

James Fenton (1983, 11)

After a very discursive investigation of a large array of individual cases of domicide, followed by two detailed case studies, we will now attempt to draw the threads together and construct a typology, if not a theory, of domicide. Domicide has been defined as the deliberate destruction of home by agencies pursuing goals, which involves planning or similar processes, and which causes suffering to those who lose their homes. For the proponents, goal-oriented planning is the outstanding characteristic, often with the public interest as a basic motive. For the victims, the salient characteristic is suffering, for those who are content to give up their homes are not victims of domicide. On these bases, this chapter will first attempt a typology of domicide, and then provide discussion of victims' responses and proponents' remediation measures.

A TYPOLOGY OF DOMICIDE

When "death" [of places] is a consequence of public policy ... the implications have to be addressed as a matter of social responsibility.

Peter Smith (1992)

Typological frameworks for understanding the nature of domicide include: spatial and temporal scales; proponents; victims; and resistance and remediation. Each of these four interrelated components is discussed separately.

Spatial and Temporal Scales

This first framework locates domicide in both space and time, paying particular attention to spatial scales. The most intimate scale of domicide is the deliberate destruction of a single home, as with the common punishment meted out to Palestinians by the Israeli army. At the scale of village or neighbourhood, the number of people remains small, and by the same token, it is easier for those responsible to divide and conquer. What is lost at this scale, as in Boston's West End, is not merely the physical manifestation of home, but also the sense of place and the meaning and social texture of community that is inextricably interwoven with the physical fabric.

When the destruction of large urban areas or whole valleys and landscapes occurs, larger numbers of people are affected, and a more complex set of land uses is involved. Destruction of home at this scale often takes place to make way for large projects such as highways, airports, reservoirs, or national parks, or, in the case of territorial homelands lost by Aboriginals, through the replacement of the original inhabitants by settlers via geopiracy. At this scale, there may be a loss of life's meaning, of belonging to a landscape or territory, which often comes about through the loss of cultural practices and livelihood. At the level of region or state, destruction may occur in war or peace, often by government decree. At this scale, domicide may be conceptual rather than physical, as in the reorganization of political space. Nonetheless, there is frequently a loss of identity, of a greater sense of belonging.

Domicide has occurred at all these scales throughout history. In war, domicide is a tool used to achieve demoralization or a punishment commonly meted out to the vanquished. In peace, both public and private development projects find that there are almost always people "in the way." Today, domicide is also ubiquitous in space, occurring on all inhabited continents and in most regions of the globe.

The sheer number of people uprooted by domicide is globally significant, probably outnumbering the 25 to 30 millions officially designated by the United Nations as refugees from political or environmental causes. Yet, little consideration is given to domicide as a global issue by governments, the United Nations, or academic studies. First, the issue has been perceived in a very fragmentary way, often in terms of isolated cases or due to a single cause, such as dam building. The second reason for the global obscurity of domicide resides in the fact that the victims rarely cross international borders; in general they remain an internal issue within individual states.

Proponents

The proponents of domicide may be considered according to their status and authority, their motives and goals, the benefits which accrue to them, and the systems they set in place to legitimize domicide. It is also interesting to consider their actions, particularly planning, in relation to theories of violence.

Two major categories of motives predominate. In war, motives include revenge, leverage against the enemy, or destruction of the physical basis of life and identity of a hated foe. Yet, the common good rationale may surface even in wartime: Hiroshima and Nagasaki were annihilated, we are told, to prevent the future losses of thousands of military personnel, including Japanese. Rather more astoundingly, American propaganda during the Vietnam War insisted that villages, regions, and even countries were being "bombed back into the Stone Age" in order to "save" them from the horrors of peacetime Communism.

In peacetime, the motives for domicide are legion and include socioeconomic improvement, environmental protection and recreation, racism or other ideology, jurisdictional efficiency, the assertion of sovereignty, the rationalization of government service provision, and the acquisition of land for settlement purposes. These general motives are manifested as specific goals, such as urban renewal, the siting of public facilities, or colonization. Such goals are almost always justified by public interest or common good rhetoric, varying from the improvement of living conditions for the poor to improvements in economy and lifestyle for much larger segments of a population, as with the benefits expected from the building of a super-dam. This type of rationale is of a different order than that used by private corporations, the primary motive of which is profit.

Those who espouse such motives and pursue such goals are not ordinary citizens. Domicide is rarely a grassroots policy. Rather, it is typically instigated and carried out by powerful elites. With a background of general motivation (e.g., increased bureaucratic efficiency or the betterment of the poor) and a set of specific goals in mind (e.g., slum clearance), these elites may also be major beneficiaries of domicide in terms of consolidation of position, achievement of an ideological goal, enhanced power, improved self-esteem, or even financial gain.

Such elites may be war leaders or academics (implicated in the genesis of ideas such as apartheid, Yugoslav ethnic cleansing, and the partition of British India), religious fanatics or powerful politicians, bureaucrats or business leaders, or any combination of these. Elite groups are sometimes led by one individual with tremendous power; the mission of Franklin Delano Roosevelt contrasts with the destructiveness of Bomber Harris and the megalomania of Nicolae Ceausescu.

Robert Moses (Moses 1970, Caro 1974, Berman 1982) is a case in point. New York's "planning czar" from the 1920s to the 1960s, Moses pursued a dream of modernization, predicated on consumerism, high levels of automobile use, and public access to exurban parks and beaches. Supported by General Motors and New Deal federal money, Moses began the wholesale rebuilding of New York in the 1930s, producing an interlocking network of new highways, bridges, tunnels, dams, parks and other large public works. His devotion to democracy – his autobiography was entitled *Working for the People* – did not prevent the massive destruction of neighbourhoods to achieve his ends. He famously remarked that "When you operate in an overbuilt metropolis, you have to hack your way with a meat-axe" (Caro 1974, 849). The meat-axe approach to the carcass of New York came to the fore in the building of the Cross-Bronx Expressway, which began to slice through that borough in the 1950s, destroying stable mixed-race neighbourhoods, blighting the areas left standing, and beginning the process that has turned the Bronx into today's urban wilderness.

To Moses, the only difference between inner-city road building and the construction of his famous exurban parkways was: "There are more houses in the way ... more people in the way, that's all. There's very little hardship in the thing." When opposed, he simply remarked "I'm just going to keep on building. You do the best you can to stop it" (Caro 1974, 876). Only in the 1960s did the opposition to this Ozymandian passion for destroying and rebuilding manage to bring it to a stop. "But his works still surround us, and his spirit continues to haunt our public and private lives" (Berman 1982, 294–5). Weeping for his lost neighbourhood, Berman "felt a grief that, I can see now, is endemic to modern life ... All that is solid melts into air."

Many engineers and planners enjoyed working with Moses, giving little thought to the lives they were disrupting in their joy in participating in giant undertakings, ostensibly for the general good. Such leaders may well be able to transmute values, so that, as Lively remarks of London's redevelopment in the 1980s: "Ruthless greed becomes entrepreneurial skill, opportunism becomes far sightedness and acumen. The ravished landscapes and blighted lives, incapable of testimony, slide into oblivion. Finally, the statues are erected; the bold, visionary figures arise in bronze upon their plinths." This testifies, in fact, that one basic motive may be sheer personal glory, even though the cost of that glory is "a landscape of secret carnage" (Lively 1991, 196)

Despite much criticism, the concept of power elites appears to hold water (Catanese 1984). It posits a loose grouping of social and economic notables who maintain close links with politicians at several levels of government and also with powerful bureaucrats. This certainly appeared to be the case in the domicide of the village of Howdendyke (Porteous 1989) and has, at the other extreme, been found at the national level in the United States (Domhoff 1978). Catanese (1984) believes that such special interest groups can readily countermand public-participation processes in politics or planning, pushing through their own goals that may or may not coincide with the public interest.

Indeed, so great was the polarization of Western societies in the late twentieth century that elite goals came to differ quite radically from those of the population at large. For example, in 1996, the Canadian federal government's Policy Research Committee asked Canadians to state their top ten goals or values (Canadian Centre for Policy Alternatives *Monitor* September 1998, 8). The general public's three chief goals were freedom, a clean environment, and a healthy population. For business and financial elites, however, these goals ranked only seventh, ninth, and tenth. The elite's chief goals were competitiveness, integrity, and minimal government. In the last decades of the twentieth century, government power was indeed reduced drastically in several Western nations, thus leaving corporations more free to change landscapes at will.

The leaders of projects are also interested in satisfying their own needs. Maslow's hierarchy of needs includes the category of "esteem needs," which involve the desire for mastery, competence, or confidence to face the world. Less positively, Maslow (Lowry 1973, 90) notes that esteem needs may take the form of a desire for reputation or prestige: "status, dominance, recognition, attention, importance." It is likely that such desires are common among domicide's perpetrators, whose actions significantly weaken the competence and confidence of their victims. More recent psychological investigations emphasize that leaders' needs include "the need to please one's ... allies. The need to appear 'macho' ... The need to capture the attention of the world ... The need to prove to oneself that one is brave, decisive, confident, and powerful. The need to avoid feelings of shame, weakness, or cowardice. The need to go down in history as a great leader" (Dixon 1987, 69).

Such needs may lead to "disastrous decisions," justified by the adage "it's for your own good" (Dixon 1987, 137). However, the factor that "has probably resulted in more human misery than all the others put together" is, simply, that the very personality traits that take people to the top and establish them as powerful decision makers tend to include the most unpleasant of human personality characteristics: "extremes of egocentrism, insincerity, dishonesty, corruptibility, cynicism, and, on occasions, ruthless murderous hostility towards anyone who threatens their position" (Dixon in Ptolemy 1994, 20). Dixon concludes: "Even worse, if that is possible, than the traits which take them to the top are those which they acquire on arrival – pomposity, paranoia, and mega-lomaniac delusions of grandeur."

This is strong stuff that may be taken with a pinch of salt, but there is sufficient evidence of a "command-and-control" approach in all kinds of planning (Burton 1994, 15) that we would do well to look with initial suspicion, at least, on the pronouncements and practices of elites. It is quite clear, for example, that business interests can subvert the planning process at individual sites (Porteous 1995), while Gleeson (1994) has shown that recent market-oriented legislation in New Zealand effectually turns consent for new projects into a commodity that can be purchased. As Gleeson (1994, 44) asks: "when money talks, who is silenced?"

The will of these individuals or elites is frequently carried out through the creation of a special authority, such as the Tennessee Valley Authority, and frequently succeeds through the good offices of the "evangelistic bureaucrat," in which "planners are the most highly developed form of the evangelistic bureaucrat" (Davies 1972, 110). Planning has been seen as a "very ruthless bargaining process" (Davies 1972, 111), in which planners tend to have the upper hand because of their belief that they have the right to specify the future and that they are capable of defining what the public interest is. These are telling claims for a profession so dependent upon the use of statutory powers and sanctions, such as eminent domain, in the pursuit of its goals.

In the liberal capitalist world, in which the common good is still an important rhetorical trope, the problem has always been who shall define it. Whether as the implementer of elite plans or as part of the elite, the evangelistic bureaucrat "legitimates his schemes not by reference to the *actual* consumer, but either in terms of his own self-pro-claimed and self-induced charisma or by reference to a range of puta-tive consumers whose wishes and wants he himself can, in impunity, define" (Davies 1972, 3). In authoritarian states, and in particular in pre-1991 communist ones, planners could achieve godlike status. As the Hungarian novelist George Konrad writes in *The City Builder* (1987, 89–90): "If God, whose other name is Plan, resides in man, then the planner is the most man-like man," and, "I plan; therefore I am. I feel my way in the world with plans. With each line drawn on the blueprint, I cut through the face of doubt."

In domicide, the elite vision or the bureaucratic plan are often supported by a law, decree, or regulation that justifies the actions taken (the *Group Areas Act*, the *Resettlement Act*, etc.). Frequently, some form of common good rhetoric is used to explain vision, law, plan, and process to the world at large, and possibly even to the victims ("these people are being liberated," or "it is sometimes necessary for people to be moved for their own good"). Thus, those who reject relocation can be defined as infantile, people simply unable to understand what is best for them. Victims, indeed, may be looked on with contempt simply because they are victims.

Further, communication between officials and relocatees is rarely clear. Some officials may be openly contemptuous of the "sentimentality" of people who wish to stay in what are "objectively poor conditions." Even those with sympathy for relocatees, who advocate social impact analysis and "planning for people as though they really mattered" (Lichfield 1996, 234) are bound up in a web of technocratic language, models, and systems that are not only incomprehensible, but also very daunting to the average citizen. Relocatees cannot always understand even the goals or motives of the proponents, thus increasing their sense of powerlessness. Their inability to formulate succinct ripostes in the language of planning may exacerbate their sense of worthlessness. Confusion is further created by the bureaucratic propensity for acronyms and the use of a host of euphemisms in the domicide process. For example, the inhabitants of "designated outports" are "relocated" by the "Ministry of BAD" according to the "Centralization Program," which will resettle them in "Approved Land Assembly Areas." Deportation, murder, rape, and the destruction of homes and property in Bosnia, for example, is euphemistically termed "ethnic cleansing," just as Cold War American governments calmly proposed the annihilation of Europe via "multiply-targeted re-entry vehicles" in a "limited nuclear engagement."

Finally, the role of elites and planners can be viewed in terms of theories of violence. Violence is defined in therapeutic situations as the crossing of boundaries without permission. Failure to ask permission is a crucial issue, especially when not only personal, but also threshold, community, and territorial boundaries are arbitrarily crossed, as in domicide. Violence demands submission, which means giving way or one's power to something beyond oneself. Violence, therefore, creates victims – people in someone's power or under someone's control. Those who perpetrate violence generally dismiss the feelings of their victims or may even ignore them altogether except as obstacles to be eliminated. In such cases, those who are victimized feel that, although hurt, they are "not seen." Much of the therapeutic discourse of victimology applies directly to the victims of domicide, although the latter often suffer collectively as well as individually (Walklate 1989). According to Kastenbaum and Aisenberg (1976, 291–4; see also DeSpelder and Strickland 1987, 384), eight factors favour violence between people: physical or psychological separation; the possibility of justification; perception of people as "the other," or as less than human; an obvious means of escaping responsibility; self-perceptions of worthlessness; reduced self-control; a situation of forced hasty decision; and anything that encourages the perpetrator to feel above or outside the law.

It is clear that some of these "psychic manoeuvres" have relevance to those who plan removal and resettlement projects. Elite groups may feel themselves, if not outside the law, then as instruments of the law or as supported by a legal system that is basically theirs. Their position, education, and power psychologically separate them from the people they plan to remove. Porteous (1992b) found that the planners and the plannees thought and expressed themselves so differently that they could be considered as two separate universes of discourse. Physically, the planners and plannees are separated by living and work environments and physical distances that mirror social distance. Consequently, those to be removed are often considered "less than," as statistics, or, at best, in a paternalist manner. Removals are facilitated when domicide can be defined as something else by the proponents, which is where common good and public interest rhetoric prove so useful. Some proponents may invoke lack of personal responsibility: "we are merely following the plan." Finally, in some cases, hasty planning decisions have to be made, so that deliberation is compromised and scruples may be abandoned.

Although we do not wish to press the analogy between planning and violence too far, it remains true that, in extreme domicide, in particular, violence is an inherent component of domicide. In everyday domicide, it is sometimes the case that what is perceived as rational, socially uplifting action on the part of the planners is experienced and defined as violence on the part of the victims. Professional planners, of course, have the invidious task of being "in the middle," attempting to mediate between the visions of political or corporate elites and the outrage of those impacted by such visions.

Victims

The victims of domicide are considered in terms of their vulnerability, the grief syndrome, and the effects of domicide upon them. Particular attention is paid to the role of memory and the possibility of memoricide.

Vulnerability.   Paradoxically, the victims of domicide who suffer quite considerably may be those who have been especially designated as beneficiaries, as in slum clearance programs. More often, they are merely "people in the way" who must, unfortunately, be sacrificed for the common good. In war, of course, they are the enemy, as specified by the elites who have decided to commence hostilities.

In all these roles, the victims of domicide become "the other," defined very often by some trait that marks them off and sets them apart from the elites, the planners, or the public at large – "those who count." They may be "the enemy," "the lumpen proletariat," "ethnics," "the working class," "ghetto dwellers," "coloureds" (in South Africa), or "bush people" (in Africa). A leading characteristic of domicide victims, then, is their "otherness;" they are people who are usually poorer, less literate, less educated, less professional, and less able to see "the global view" than the elites, the planners, or even the public "who count." They are also likely to be significantly disempowered, lacking the resources, understanding, vision, or experience to combat the plan that is to devour them. In other words, they are often people who are used to being, and perhaps expect to be, "kicked around" by "them."

Such characteristics may be found among inner-city tenants, who have no property claims on their dwellings, or workers who are part of a paternalist system of company towns based on industry or resource extraction. If the victims are rural or Aboriginal, it is often assumed that their way of life or culture has little value, and that the next generation would not wish to live in such a way or in such a place. In the creation of airports and highways, the definition of victim is not so clear. It certainly expands to include middle-class people occasionally, such as those living in a peripheral "stockbroker belt" required for an orbital freeway. Most often, however, the victims are people who "may not think of their bedroom, their corner shop, or their garment factory as an inner city development site" (Lively 1991, 60–6), people who yet may be in tune with "the power of the place, its resonances, its charge of life, its coded narrative." Above all, domicide victims are generally subordinate – people without authority.

Nevertheless, victims vary tremendously in their level of vulnerability. The places most vulnerable to domicide are inner-city areas, urban fringe zones, dammable valleys, remote islands, and the so-called *terra nullius* (empty regions) occupied by scattered nomads. The people most vulnerable to domicide, as already noted, are those most likely to be considered as "others" by elites and planners, people who are likely to be able to mount least resistance to the project. "Otherness" appears to reach its extreme among Aboriginal peoples or those who live traditional lives. This is true whether we are considering those impacted by Western planners or by the elites of India, Africa, or Southeast Asia. For example, although only 8 per cent of India's people are classified as "tribals," tribal people make up 40 per cent of those who have been displaced by dams, military installations, opencast mines, and other grandiose installations (French 1998). The extremes of vulnerability, then, are to be found in extensive desert, prairie, or forest zones doomed to be "used" or "settled" from a metropolitan centre, or remote islands valuable for their resources or isolation, the latter being of uncommon interest to the military.

The characteristic feature of most of these vulnerable places and peoples is their remoteness or isolation. They are either physically remote, as in equatorial forests, or they are socially remote from centres of power that may be physically close by, as in poor inner-city neighbourhoods razed by entrepreneurs who can look down upon them from their office towers. Islands, of course, are quintessentially remote in both ways. And, above all, the victims of domicide are almost always smaller in numbers than those destined to benefit from the common good. Small in number and relatively powerless – their chances for successful resistance are not good.

The Grief Syndrome. The response to domicide, and the way people are affected by it, also varies a great deal. Those who are able to deal with change easily, see socio-economic advantage, or agree with common good rhetoric, may benefit. For them, by definition, the change they have undergone cannot be called domicide. Earlier, we saw cases in which relocation had positive results. Nevertheless, victims often suffer through years of uncertainty before they lose their homes. They have to stand by helplessly while a sort of "planning blight" sets in, destroying the value of their homes, and then they must adapt to their new homes through a significant process of change. Others must weather rejection by the community to which they have been transplanted.

The many case studies of uprooting in Coelho and Ahmed (1980) confirm that sudden change of any kind is a source of stress. Forced relocation, often carried out abruptly, creates an enormous amount of stress, which is sometimes manifested in socially deviant ways. Trimble opines that "outcomes of forced migration, the type where people are compelled to leave an area when their presence is not desired by an external agent, are disruptive, occasionally tragic, and in many cases, generate irreversible problems" (Trimble 1980, 455)

Thus, for many, sometimes a majority or even a whole Aboriginal group, the loss of home may bring sadness and grief. Some mourn the loss of ancestors' graves and ancestral land, while others lose friends, extended family relationships, and local social networks. Some may suffer physiological, psychological, or social distress, and they can feel anger or a sense of helplessness. Some may accommodate their grief by focusing more strongly on kith and kin, while others will idealize their lost home. Some, particularly the elderly, may die or commit suicide.

For those who survive, there may be resentment at having to suffer for the supposed greater good of the community and against those officials who are responsible for their fate. This resentment may be translated into a long-term mistrust of government, and world views emphasizing powerlessness may emerge: "Politics knows only two basic methods: The first is cheating, the other violence" (Brink 1974, 236). Kazantzakis (1974, 86) concludes that "you're either a wolf or a lamb. If you're a lamb, you're eaten up; if a wolf, you do the eating" and, in a richly Christian spirit, the despairing "Blessed are the unmerciful; Blessed are the violent, for they shall inherit the earth."

The grief syndrome among the dispossessed is worth extended consideration. Eviction has occurred since Adam and Eve, and there is rich literary and anecdotal evidence of its psychological effects. But it was Marc Fried, employed by the Massachusetts General Hospital to survey the Boston West Enders whom its extension was about to displace, who first gave the domicide grief syndrome the imprimatur of scientific medical research.

Fried made an explicit comparison between forced relocation and bereavement, both of which generate grief reactions that can be intense, deeply felt, and sometimes overwhelming. People can grieve for a lost home as they grieve for a loved one now dead. Marris (1974, 25, 44) followed up this lead with studies in both Britain and West Africa, which confirmed that grief, apathy, and a feeling of "the aimless futility of life" can readily follow relocation. In Lagos, Nigeria, those forced from their homes complained bitterly: "It seems like being taken from happiness to misery;" "I fear it like death."

The domicide grief syndrome is clearly not restricted to Western cultures. In a review of forced resettlement across Africa, Chambers (1969, 176) found that oustees tended to adopt one of two polar attitudes to their disruptive situation. Individualism, independence, self-help, and a high degree of activity were found at one extreme, while at the other was "a high degree of inactivity and apathy, combined with dependent attitudes," all common features of grief. A similar bipolar response was found in Algeria (Sutton and Lawless 1978, 342). Marris (1974) notes that a bustle of activity and apparent insouciance may sometimes presage a delayed shock reaction, worsened by a failure to do grief work. For a few, the trauma of relocation simply cannot be borne, and they commit suicide. A frightening report from Colombia (Vidal 1997) suggests that the indigenous U'wa people, their homeland on the point of being destroyed by oil companies, have considered mass suicide.

A number of attempts have been made to codify the human response to bereavement or the confrontation of the individual with the process of dying. One of the first was the idealized model of Kübler-Ross (1992) that suggested that when confronted with loss, a person may go through stages of shock, denial, anger, guilt, bargaining, depression, resignation, and acceptance. After the predicament has been acknowledged, there is generally considerable anger, vulnerability, and dependency. Crisis resolution usually comes with acceptance, although some of the dying fight to the very end, refusing to go gently.

If grief is a person's emotional response to loss, mourning is the process of incorporating the experience of loss into our continuing lives. There are many models of the grief/mourning process (DeSpelder and Strickland 1987, 212), but almost all of them begin with "shock," go through a middle period of "grief work," "separation pain," or "experiencing the pain," and conclude with some form of "re-establishment" or "reintegration;" "grief is essential if we are to move on in life" (Walter 1990, 125). Survivors may need a variety of coping mechanisms, including rituals, support groups, and social institutions, such a grief counselling, in order to come to terms with the fact of loss.

Grieving for a lost home will likely vary from the grief that follows a human loss. First, we know human death is inevitable, yet many of us believe that our dwellings, neighbourhoods, landscapes, and valleys have inherent permanence. They are bigger than us; they are centres of stability in a rapidly changing world. Second, when individuals die, at least in Western societies, there is little thought of blaming some outside agency; as in a natural disaster, death is an unrepealable act of God. The victims of domicide, in contrast, may be well aware that their homes have been destroyed by elite groups for some purpose; there is someone to blame, if it is only "them." Third, acceptance of domicide is not like the acceptance of the death of a loved one. After the latter, life goes on; whereas those who suffer domicide may well feel that acceptance of

their loss is yet another blow to their self-esteem, their sense of empowerment, their identity. Grief counselling for domicide should not blind-
ly follow the dying-bereavement grief models of social workers nor the
strategies of helping those suffering from post-traumatic stress disorder.

In my study of the village of Howdendyke, East Yorkshire (Porteous
1989), I made an attempt to codify grief stages and compare them with
the above sociological models. Both relocatees and those permitted to
stay in the half-destroyed village displayed a variety of emotional responses including sadness, depression, and resignation. Because the domicidal
process was so slow and insidious, there was little shock but a good deal
of denial, summed up as: "but surely they won't do this to us?" After
recognition of the process had set in, the pain felt was chiefly characterized by laments for the past, wherein even people still living in the
remains of the village tended to use the past tense while speaking of it.
There was some idealization of the former village and negative comparisons with current living situations, although some acknowledged that
removal had made their lives easier in some ways. A second category of
response involved the assignment of blame. And when efforts to halt the
domicide process were seen to be of no avail – given the power of the
companies, the bureaucracy, and the politicians against a community of
200 persons – resignation and depression set in, followed by acceptance
and an attempt "to make the best of it" (Porteous 1989, 170–3).

In my work on this village I have acted as unofficial therapist for not
a few villagers, being an active listener and friend as they came to terms
with the loss of the physical and social fabric of their lives. The publication of my book was valuable in several ways. It provided support
when public meetings were held to try to bring village destruction to a
halt, and it has become a valuable prop to memory for those whose
social and physical landscape has disappeared. My own grief at the
domicide of my village has been somewhat assuaged by the time-honoured self-therapeutic practice of "writing it out." A second wave of
personal grief, however, set in in 1997 when my own former house was
finally demolished. Again, writing it out proved helpful. Here are some
of my verses that fit most closely with this book's argument:

DUE PROCESS

Council has always proceeded legally and according to due procedure.
        Mayor Robert Park, Boothferry Borough Council

        Government and Business, hand in glove,
        Destroy my village. Distanced, staying clean
        They send in hired hands. Seen from above

These managers, dull yet abstractly keen,
Are doing something useful – making jobs
In backward villages for laid-off yobs.

There's no escaping this cool juggernaut's
Bulldozing of the future from the past.
No present, this! And there, beyond all courts'
Surveillance, politicians, and bureaucrats
And company directors golf their holes,
Where nothing mars their joy nor thwarts their goals.

JOBS

Workers with jobs come to destroy my place.
Cranes and bulldozers take these hovels down,
Crush up the ancient bricks. Women, old men
And kids look on in bafflement, as their known world
Collapses. It's rubble now, the Square, where
We had Bonfires, played Mischievous Night, and
Where our elders pubcrawled in one pub,
Shopped at the Post Office, danced the Hokey Cokey
In the hall. All rubble now, and better for it
Say the managers, to be asphalted over, a
'Hard standing' for a fleet of haulage trucks.

And now, what was irregular, and bright,
And three-dimensional, and full of life,
And wrong, shrivels to a harsh black flatness
Where stand oily shallows after rain.

My place is like a dying cut-down whale
With busy men dismantling, hacking, flensing.
Don't get me wrong, such work means jobs.
Jobs are a form of cleansing.

OBSCENITY BREAKS IN

They are making a killing.
How do they measure their lives?
Endlessly more killings must be made.

Can vision enter the equation?
Unless there's vision, mind, the people perish.

Not here. Their vision's of a slab of concrete
That's growing money. Of a local habitation
Still named, but suddenly becoming memory.

THE KÜBLER-ROSS APPROACH TO DYING

Denial wasn't in the cards, so
I tried anger, it was only
Right to try to stop destruction,
Rage against unmitigated
Business murdering my village.

They were stronger, richer, harder.
Blandishments were tried, and threatening,
All of which distraction covered
Up the endless hopeless slither
Of the smothering of my village.

Now I've come to some acceptance
Yet I'll demonstrate their callous
Unforgiving work by witness,
Journalling the lies and truths a-
-bout the murder of my village:
The Kübler-Ross approach to dying.

LANDSCAPES OF THE DAFT

When all the world is concrete
And only starlings sing
We'll wonder what folk used to eat
When gardens glowed with green,
We'll speculate and excavate
To find how people lived their funny
Lives before the concrete
When all we've left is money.

In the case of Howdendyke, the two worlds of discourse of planners
and plannees never came together. What villagers saw as loved homes,
company directors derided as "hovels," and bureaucrats saw as "substandard housing."

Although most Howdendykers came to terms with their loss, some
still mourned their village. A more extreme case of grief, coupled
with severe difficulties of acceptance, can be found among both the

Christian Palestinians who fled the village of Kafr Bir'im in 1947 and their co-religionists who stayed in what was to become Israel. The tented refugee camp in Lebanon proved too difficult a life for some. "My father ... never really recovered from losing everything. He hated the tents, and he missed his village and his old life in Palestine. He would have given anything to return, but he knew that all his friends who had tried to sneak across the border to their old villages had been shot dead by the Israelis. Sometimes he would just sit there looking at the key of his house in Kafr Bir'im and the title deeds the British had given his father to prove the ownership of our land. He got ill and very depressed. It was as if something had broken inside him. He died in 1956. He was only thirty-four years old" (Dalrymple 1997, 270–5). His family had to endure several further relocations within Lebanon, as refugee camps were attacked or arbitrarily bulldozed. After forty-seven years of exile, his daughter said: "After all I have suffered from the Israelis and the Lebanese, I would like to go home even if it meant I was naked and starving ... Even if I had lived a hundred years here, I would still like to go back to Palestine, go back to Kafr Bir'im where no one can tell me that ... I don't belong" (Dalrymple 1997, 270–5).

Her compatriots who had remained in Kafr Bir'im and been given Israeli citizenship had come to a greater acceptance of their lot: to be relocated away from their village and to have to ask permission to visit their old church there. At the relocation, the older relocatees "cried for many days because they had lost their homes and their land," but were reassured by the Israelis that they would some day be allowed to return. After much effort, in 1953, the villagers won a court case that declared their eviction unjust, upon which the Israeli army declared the area a military zone and destroyed Kafr Bir'im by aerial bombardment: "All the villagers went up onto that hill and watched the bombing of their homes. They call it the Crying Hill now, because everyone from Kafr Bir'im wept that day" (Dalrymple 1997, 367–9). Later, some of the village land was given to a kibbutz.

The villagers regret their "wonderful village" but "could do nothing," and "still feel betrayed." They are most resentful of the attempted memoricide of the village: the site is now part of a national park, but the park interpretation material makes no mention of the Christian village that recently existed there: "We've been edited out of history ... they don't admit to our existence. Or to the existence of our fathers and grandfathers and great-grandfathers." Though the former life is idealized, the blow to memory and to identity is bitterly felt. Today, the former Palestinian villagers of Kafr Bir'im must pay a national park fee to view the remains of their village, yet: "When we come here, we are happy ... All the old memories come back. We remember many things: the streets, the homes, the neighbours. Everything" (Dalrymple 1997, 372).

Memory and Memoricide.   The case of Kafr Bir'im brings up the next element of our assessment of the victims of domicide, the issue of memory. Memory cannot be utterly expunged while there are rememberers who can pass on stories to future generations, but it can be mortally wounded when these stories cannot be backed up by accessible documents or physical structures on the ground. Those whose task is to destroy memory, then, will take care to expunge such physical structures.

Memory is important at the national level and becomes part of nationalist myth; Serbian mythic memories of the Battle of Kosovo in 1389 are part of the background to the wars in the former Yugoslavia in the 1990s, while equally questionable myths refuel the "temple wars" of India, resulting in the destruction of an Ayodhya mosque in 1992. History is hijacked by the winners, who promulgate their useful myths (Swift 1993). In terms of domicide, myths useful to the proponents include "the certainty of progress," "national destiny," and a world organized by the powerful "from the top down." Those who resist these views are likely to fall into mythic traps such as "the good old days" or "the nobility of the oppressed." As Orwell emphasized, history is constantly rewritten, from official Chinese history books to American theme parks.

The physical destruction of place and artifact is commonly used in attempts to erase memory and alter cultures. We are familiar with the wholesale destruction of monasteries and of Tibetan buildings in Lhasa by the Chinese after 1959. The Serbian destruction of mosques, graveyards, libraries, and archives in the early 1990s was an attempt to erase the Bosnian Muslim imprint from the land. In Israel, former Palestinian villages are literally rubbed out, only to re-emerge in outline as the old villages' deep-rooted cactus hedges "keep sprouting again and again to mark the sites of the former garden boundaries and the shadows of former fields" (Dalrymple 1997, 863).

The example of Kafr Bir'im, noted above, is only one case of what appears to be a systematic attempt by Muslims and Jews to remove Christian structures from the landscape of the Middle East (Dalrymple 1997). Turkish archaeology ignores the Armenian heritage, and semi-official destruction of Armenian artifacts continues in a deliberate campaign to destroy all evidence of the long presence of Armenians in eastern Anatolia. Similar efforts are being made in Israel. In the Old City of Jerusalem, all Arabs living in the Jewish Quarter were evicted in

1967. Since then, radical settler groups have been buying up properties in the Muslim, Christian, and Armenian quarters, in which non-Jews find building permits difficult to obtain. There appears to be a concerted attempt to Judaize the Old City, the long-term results of which will be religious and cultural suppression and the weakening of historical memory (Said 1994, 189).

Individuals combat such attempts to expunge memory in a variety of ways. While authorities may try to preserve major artifacts and archives, ordinary people rely on photographs, letters, small mementos, and the retelling of their experiences. Peasants carry away a small bag of soil from their ancestral lands. Although they may have suffered much loss – of property rights, dwellings, land, and social networks – the relocated can maintain both their identity and their attachment to lost homes in this way for three or more generations. To paraphrase McKie (1973, 15), the triumph of the rational and efficient over the picturesque and sentimental is rarely complete.

The popularity of autobiography, childhood novels, and family history research testifies to the importance of memory in our lives. The book *Planned to Death* (Porteous 1989) has become Howdendyke's "memorial volume," and it has generated a number of letters from readers who credit it with the revival within them of buried memories. Andre Brink's novel *Looking on Darkness* (1974) helps preserve the memory of Cape Town's District Six, while the work of Evenden and Anderson (1972) demonstrates that Japanese-Canadians evicted and interned during the Second World War strive to retain memories of their internment camps, which have become an integral part of their collective identity. Whereas Salman Rushdie's theory of fiction demands the chewing over of "big chunks of the universe," Julian Evans (1993) rightly suggests that most of us live lives of detail at the local level, so that the parochial "tiny pieces of human experience" that Rushdie disdains are, in fact, just the landscapes of the heart that are so important to us.

It is interesting to note that new trends in medical thought agree with this proposition. Both allopathic (McCarron et al. 1994) and alternative (Drum 1998) medical traditions are beginning to recognize that health and well-being are substantially contextual, and that a significant part of that contextuality involves belongingness and feelings of "being at home."

The final outlet for the victims of domicide is resistance, while for the proponents, the issue of domicide is addressed through processes of remediation. These issues of social justice, which are discussed in the next section, are more relevant to everyday domicide than to the extreme variety.

Too often, in the past, the victims of domicide have been ignored. Their moral right to resist and to enjoy remediation is forcibly asserted by planner Peter Marris.

Resistance and Remediation

When those who have power to manipulate changes act as if they have only to explain, and when their explanations are not at once accepted, shrug off opposition as ignorance or prejudice, they express a profound contempt for the meaning of lives other than their own. For the reformers have already assimilated these changes to their purposes, and worked out a reformulation which makes sense to them, perhaps through months or years of analysis and debate. If they deny others the chance to do the same, they treat them as puppets dangling by the threads of their own conceptions.

Marris (1974, 166)

Domicide is frequently protested, but when resistance occurs, it is often ineffectual. Attempts to soften or channel resistance, coupled with the laudable motive of allowing plannees to have their say, generally take the form of some type of public participation exercise. The goals of public participation in planning are to strengthen the influence of both citizens and planners in decision making vis-à-vis economic and political elites, and to reduce differences in power (Fagence 1977, 46). Typologies of citizen participation include that of Milbraith (1965), who divided the behaviour patterns of political activity into gladiatorial (holding public office), transitional (attending a public meeting), spectator (voting), and apathetic inactivity. Spiegel and Mittenthal (1968) suggested seven elements including: information, consultation, registration, shared policy and decision making, joint planning, delegation of planning responsibility, and neighbourhood control. This concept of an ascending order of popular control of change has been elaborated by Arnstein (1969) and Lipsky (1970).

For Arnstein, citizen participation is another term for power. She suggests a framework that arranges citizen participation efforts on a ladder, with each rung corresponding to the extent of a citizen's power in determining a plan or program. The first two rungs – manipulation and therapy – are basically non-participatory whereby persons in power "educate" or "cure" the participants. At rungs three, four, and five, citizens hear or are heard but cannot ensure that their views will make a difference. At the upper rungs, people are able to negotiate with persons in power and may, at the highest level, hold decision-making seats or managerial power.

Arnstein's ladder is interpreted by Fagence (1977) to mean that there are varying degrees of participation tolerated by plan-making agencies or varying degrees of involvement accepted by participants. However, there is another way of seeing this framework in relation to prospective victims of domicide. In reviewing the case studies of chapters 3 and 4, there appears to be a correlation between levels of resistance and the degree of power achieved by persons who resist the threat of domicide (Lipsky 1970). There are three broad categories that describe resistance: little or no resistance, limited resistance, and major resistance. Each of these, in turn, appears to relate to the relevant rungs on Arnstein's ladder. For example, when there was major resistance, a degree of citizen control was achieved (airport construction was stopped or a more acceptable project was built). When there was virtually no participation in decisions, affected persons were manipulated successfully by power holders. Having recognized this correlation, it must nevertheless be noted that it is possible that the choice of citizen participation measures could have some effect on the amount of resistance encountered. When well designed, public participation measures might diminish resistance.

At the bottom rung of Arnstein's ladder, little or no resistance occurs when project developers do not inform people that they are to lose their homes until it is too late for them to take action. In response, citizens adapt to the fact that their community is dying or simply are unable to make any decision. The villagers of Howdendyke illustrate this circumstance. Although the agenda to expand industrial holdings had been clear to industry and planning authorities since 1968, it was not until eleven years later that the residents of Howdendyke, East Yorkshire, learned of their fate when permission for improvement to a cottage in the village was denied (Porteous 1989, 160). Their reaction to the planned death of their village was mixed. Some were indifferent, while others believed that such a thing could not happen. Some attempted to sell their houses, while others tried to change the situation through letters to the planners, politicians, and the editor of the local newspaper. But few had the knowledge of planning procedures or the ability to organize a resistance that would make any difference. A survey in June 1981 found that most Howdendykers, of whom two-thirds were tenants, wanted to stay. All the same, many of the villagers moved away. For the less than one-third who are left, there has been a recent change worthy of note. Some of the remaining houses have had small improvements made to them by the industry that owns them. It is possible that due to Porteous' work, the company has developed a conscience, but if so, it has come too little and late for most people.

Limited resistance to the threat of domicide is expressed in a number of ways and frequently occurs when there is some degree of involvement of the victims in their ultimate fate. This involvement is often pure tokenism but may result from situations when planning for resettlement is undertaken. In the Volta River project, despite significant preplanning, some resistance was encountered when people did not buy into the "self-help" programs designed to help them create new homes. Similarly, with the construction of the Kariba Gorge Dam, preplanning occurred, but the Gwembe Tonga resisted settlement for the following reasons: the program threatened people's basic securities; they did not understand the technical facts; the project was the result of an order from outside their settlement; and they were made to suffer for the longer-term good of their community.

When major resistance occurs, a project may be cancelled or redesigned, and in the latter case, there will be significant involvement of affected persons. This resistance is likely to be successful because social networks are well developed or develop quickly based on strong horizontal (relationship between a community's social units) and vertical (relationship between community's social units and outside political, social, or economic institutions) integration (Rohe and Mouw 1991, 58, 59). Two examples of major resistance that resulted in quite different end points include the resistance of the people of Cublington in England against the third London airport and resistance on behalf of the residents of Crest Street, Durham against construction of an expressway.

Some of the most massive resistance efforts have been mounted against dam building. These will be discussed in chapter 7, which outlines proposals for combating or mitigating domicide. At this point, the brief discussion of participation and resistance above underlines the need for better planning practice in dealing with domicide victims as well as the necessity to investigate the shortcomings of current practice.

While there are circumstances in which citizen involvement has changed the course of decision making, earlier chapters have described many situations in which this did not occur. In these circumstances, a number of measures have been used to involve and provide mitigation and/or compensation for the victims of domicide, ranging from the proactive (public meetings and hearings and social impact assessment) to active intervention (planning), and, finally, to measures to be used as a last resort, such as expropriation.

Public Meetings and Hearings. Public meetings and hearings were widely used in the 1960s to gather public input on the effect of major projects, and, in the process, the need for assistance to the victims of domicide was identified. This practice continues today since it provides a number of benefits to project proponents, including: legitimacy due to legal authority and frequency of use; the appearance of an open process; an inexpensive format (at least before the advent of intervener funding, which allows people who would otherwise not be able to attend to participate); and, finally, the opportunity to provide information about a project and, in turn, gauge public attitude and receive public comment.

However, the process is frequently dominated by organized groups and agencies that have their own agenda and neglect the broader social views. It is often an arena for confrontation and/or intimidation, and it is used as a ritual or to co-opt participants. Frequently, it is limited in terms of time and scope of the discussion. For example, the effects of a development are discussed, rather than whether the development should occur or not (Reed 1984, 13–18). For the individual, there are significant drawbacks; the process can be mystifying and costly both in terms of the time necessary to participate and in the emotional toll of such participation.

Reed (1984, 18–19, 51–2) identified seven criteria for a fair hearing on environmental issues. These criteria are generally applicable to the situation of probable domicide: all members of the public should have the right to appear; participation should be enabled early on in the project before irrevocable decisions are made; sufficient notice of hearings and other procedures must be given; an impartial board should preside over hearings; participants should have easy access to all relevant information and government expertise; participants should be provided with research time and financial aid according to predetermined criteria; and complementary techniques for public education and comment should be provided.

Meetings and hearings are now frequently augmented by the use of social and environmental impact assessment and planning processes.

Social Impact Assessment. Social impact assessment has emerged as an essential component of environmental impact assessment. The primary focus is on estimating and evaluating the social change that would be brought about by specified project alternatives. Based on this assessment, the project can be altered, or mitigation or compensation can be prescribed.

The primary focus of these assessments is at the level of community (Hindmarsh et al. 1988), with especial interest in the economic effect on the community (Bowles 1981). Methods have been gleaned from the social sciences and have included the Delphi technique, systems approaches, trend extrapolation, and quantitative modelling (Soderstrom

1981, vi). In addition, social indicators are frequently used to permit quantitative measures of the condition of a group of individuals over time. Such measures include population density, mobility, housing, crowding, transportation, desirable community growth, and community cohesion. They may be augmented by a consideration of the perceived need for property acquisition, cultural heritage designation (Weiler 1980), or landscape evaluation. The gathering of these indicators is enhanced by the use of public participation during the assessment process. Hyman and Stiftel (1988, 41ff.) suggest that public participation allows a more direct and actual role through advisory committees, public meetings, or group process techniques. In addition, they suggest that participation should permit better design of mitigative measures, gain support from diverse groups, and act to gain people's confidence. Finally, they acknowledge the need for public participation to be combined with an extensive media and outreach program.

Yet a review of a number of sources on social impact assessment (Bowles 1981, Clark and Herington 1988, Finsterbusch 1980, 1983, Hyman and Stiftel 1988, Krawetz 1991, Lang and Armour 1981, Lattey 1980, Soderstrom 1981, Weiler 1980, and Wolf 1981) demonstrates that there is little focus on the individual and home. Undoubtedly this results from two factors: the apparent need to provide quantitative analysis and the fact that most social impact assessment is undertaken on behalf of the project proponent. Only Finsterbusch (1980, 103–5) recognizes the significance of attachment to home, particularly the significance of home ownership as part of the "American dream," and the relationship between home value and neighbourhood satisfaction. While he suggests that the main function of housing is the provision of living space, he also acknowledges the significance of home in terms of attachment, security, and concept of self. However, no guidance is given as to how to take these values into account.

Lang and Armour (1981, 47ff.) identify a number of other problems with social impact assessment beyond the lack of focus on the individual. These are the following:

• Social indicators are determined mainly by the project proponent.
• Assessment panels and the public are often unable to assess the significance of projected changes.
• Public input is often limited, and no opportunity is given to comment on indicators to be collected or the process to be undertaken.
• Not enough consideration is given to "way of life."
• Too little information is given to those most affected by the project.

- There is uncertain follow-up on social impacts. Simply requesting "careful planning" in dealing with a community does not mean it is going to happen.

In summary, social impact assessment continues to strive to provide a means by which social and economic consequences arising from project development can be mitigated. Despite practical and academic attention, it has failed to gain the acceptance enjoyed by environmental impact assessment, which has had more success in describing impacts. While much energy has been put into the development of precise descriptors in the form of social indicators, these do not appear to be easily interpreted by assessment panels.

Planning Processes. Planning processes are very commonly used to provide mitigation. As discussed by Lawrence (1992, 23), there are significant similarities between the planner and someone who undertakes environmental and social impact assessment. They share a focus on something that will eventually be built, although the planner's range of concerns is greater, touching on social, economic, and cultural factors. They share a reliance on the rational planning model and have both struggled with the role of public participation. They both rely on the production of a plan or impact statement, and, in both cases, the planners and assessors are seen as advisors and facilitators. However, planners manage change, whereas assessors simply evaluate proposed changes. Further, planning has for many years concerned itself with broader notions of public interest and an emphasis on realizing a positive end, while environmental impact assessment (in theory undertaken to assess whether a project should be built) tends to start from the perception that there is an imbalance (proposed destruction of the environment) to be righted.

In the context of everyday projects described in this book, planners are seen primarily as managers of change on behalf of a commission or government agency that is carrying out a large project. In this context, also, planners are inevitably subject to the approval of politicians or corporate elites, who have a particular definition of the public interest (Porteous 1977, 316–17). Nevertheless there are examples of attempts being made to accommodate the needs of the plannees. Two are explored here.

Finsterbusch (1980, 129–35) studied the relocation of Hill, New Hampshire, by the American Army Corps of Engineers in 1940. The town's residents, under the leadership of their state planning director, relocated to a new, carefully planned town, which kept most of the community together. According to Finsterbusch, the success of this project relied on: leadership, consensus, co-operation, and community solidarity; the existence of industries that were willing to relocate; the existence of an attractive and affordable site for the community; and, finally, the presence of financial assistance to the homeowners and the community. Overall, "community cohesion, identity, and pride were also enhanced" during the relocation process.

The second example is the Volta River Project resettlement undertaken in Ghana in the 1960s and planned by a Preparatory Commission sponsored by the British and Ghanaian governments. While this project was undertaken a generation ago, it serves as an example here because it was constructed at approximately the same time as the Columbia River projects analysed in chapter 5 and since it was the subject of a detailed critical review. The project goal, to leave no one worse off than before the creation of the reservoir, was to be achieved through a combination of compensation and resettlement through self-help and incentives. Information to support the project goal was collected through an extensive social survey. Self-help was seen as "an excellent way of rehabilitating" the victims of domicide (Kalitsi 1970, 40). In addition, all public sector infrastructure was to be replaced, and private interests were to be compensated at market value plus a "disturbance element" of 20 per cent of the assessed value of private buildings.

Villagers were given an opportunity to choose the location of their future home. Difficulties arose only when people did not understand the reason for one choice of location or another or when people were unwilling to leave the traditional lands of their ancestors. Kalitsi's (1970, 54) description of the evacuation of these areas gives some insight into the efforts that were made to provide assistance to the evacuees: "On the day of the evacuation, a despatch team went with the transport, encouraged the people in packing up and loading, and issued each family head with a householder's identification card. Social workers travelled with the evacuees on the boats and lorries. In some cases, the journeys took as long as three days, and there were difficulties created by inclement weather ... On arrival, a reception team of social workers with predetermined house allocation plans conducted the people to their houses, issued them with rations from food donated by the World Food Program, and stayed with them to assist them to find their feet" (Kalitsi 1970, 54).

That the evacuation went as smoothly as it did was attributed to the presence of social workers in the field for a number of years before the final move, participation by villagers in the choice of future location, and the involvement of community leaders. Payment in the form of housing materials assisted resettlement, as did the provision of simple village layouts, house plans, necessary machinery, and the construction of public amenities. While the development of small communities, different types of housing sensitive to the regions, and the use of communal labour were intended, the short time frame that was eventually faced by the project forced the use of imported labour, uniform housing design, and the creation of small towns.

Advocacy Planning. Advocacy planning has been less frequently used. The citizen advocate is "an expert [used] to confront an expert, to enlist on behalf of citizens the services and resources which help to equalize the struggle against the system" (Stinson 1974, 39). In order to do this, the notion of value-free planning is rejected, and planners were to become advocates for "what they deemed proper" (Davidoff 1965, 331–2). Planners were to rely on a person's life experience in the area and on their ability to measure and observe, rather than on technical evaluations. Everyone was to be given an opportunity to provide his or her opinion (Breitbart and Peet 1973, 97). Two types of advocacy evolved from this position: direct representation of a specific client group on a planning issue and indirect pressure on behalf of a community without being tied to any particular interest. While the use of advocacy planning was initially hailed, there were also a number of problems, described by Breitbart and Peet (1973, 101), including: the impact of an advocate's personal values and biases on a community group (sometimes taking the form of manipulation); the exclusion of citizens from the planning process; and data manipulation. Advocacy planning was also considered chiefly as a reaction to crisis.

Expropriation. In sharp contrast, expropriation is the state's action of taking or modifying the property rights of individuals through the exercise of sovereignty. If no compensation is paid, such an action amounts to confiscation. Where compensation is made, the terms "condemnation" or "eminent domain" are used in the United States to denote the compulsory acquisition of private property for public use or benefit; compensation should be "just." Abrams (1971, 100) notes that the exercise of eminent domain in the United States has been much more widespread than the equivalent "compulsory purchase" in England, because "people lacked the deep attachment to site that was characteristic of Europe." In Canada, Todd both defines expropriation and recognizes its chief deficiency: "In general terms, 'expropriation' is the compulsory (i.e., against the wishes of the owner) acquisition of property, usually real property, by the Crown or by one of its authorized agencies. The power of expropriation is generally recognized as a necessary adjunct of modern government, but its exercise nearly always results in a traumatic experience for the affected property owner" (Todd 1992, 1). This definition is obviously germane in the study of domicide.

In commenting on the subject of expropriation and compensation, Knetsch (1983, iv) highlights the dilemma of "why some interests might be favoured over others." In particular, he recognizes that there are values that do not have a direct expression through market exchanges and thus receive less recognition. However, he suggests that market value has been chosen because values like emotional attachment or sentiment are not measurable, that attempting to reach such values could lead to different compensations for similar properties, that such values might lead to "excessive" claims, and, finally, that there is a long-standing legal principle that an owner must give up land required by "the community." He further suggests that as an alternative to compensation, every effort should be made to exchange one property for another or to make offers in excess of the market price. (p. 48)

Knetsch's views are augmented by a study of the perception of property settlement payments undertaken by Korsching et al. (1980). In general, they found that most property owners felt that they were not paid enough despite the fact that they received greater than the appraised value. Their antagonism was directed toward the project developers and their representatives and the method by which the latter undertook appraisals. The tools of expropriation and compensation, even when used when all else fails, thus leave much to be desired from the viewpoint of the victim. If anything, they highlight the importance of proactive measures to compensate the victims of domicide.

CONCLUSIONS

We must ... mobilize ourselves and struggle to preserve the remembered sense of community and integrate it into future attempts at change.

Gallaher and Padfield (1980, 22)

This chapter began by providing an initial typology of domicide, which essentially pitted powerful elites and bureaucracies, intent upon their goals, against relatively powerless victims who happen to be "in the way" of desired change. Elites and victims interact in two major ways: by protest and resistance on the part of the victims, and by remediation of some kind on the part of the proponents.

Few attempts at resistance have been successful. When the response to domicide is examined, it is found to range from major resistance and confrontation to an almost complete and silent acceptance. Major resistance may result in some degree of citizen power being achieved. Limited resistance may at least result in the victims being informed, placated, or consulted. In each of these, some form of partnership with the victims may occur. There are, however, many circumstances in which there is no participation by the victims, and, therefore, they are merely manipulated to achieve the ends of those in power.

The range of response suggests that processes are needed to ensure that useless confrontation, passive acceptance, or tokenism do not occur. This chapter therefore assessed the five techniques that have been most commonly used. Public meetings and hearings have been frequently used in the past and continue in common use today. For the project proponent, they provide an efficient tool in terms of time and cost. For those affected, the public hearing is the least sympathetic means by which their loss can be measured and hence compensated. Social impact assessment is the most well known anticipatory means used to measure the value of what is to be lost, but this measure has failed to develop the same utility as environmental assessment. Planning processes have been used to provide mitigation, particularly through the creation of physical infrastructure to replace what is lost when domicide occurs. Success in such planning projects most often relies on leadership, consensus, and community solidarity. Advocacy planning is also useful, but this tool is not frequently used at present. The least successful means of determining compensation is expropriation. While many changes have been made to the legislation that governs expropriation in recent years, it must be recognized as a mechanism of last resort. To criticize expropriation for not accommodating intrinsic values is to misunderstand its intent. There will always be times when a monetary value of loss is required, and this need points to the developing field of multiple account analysis.

Each of the traditional measures described above could be improved through greater acknowledgment and respect for the lives of others. Indeed, the epigraphs to the major sections of this chapter summarize two of its main themes: that the key to successfully assisting the victims of domicide lies in the respect that is paid to the meaning of their lives and their sense of community; and that any processes designed to help them must come to terms with the importance of intangible values, values that as yet defy quantitative measurement. An attempt to suggest such processes will be made in the following chapter.

CHAPTER SEVEN

Ending Domicide?

In the vocabulary of profit, there is no word for "pity."
John Updike (1978, 241)

Where, after all, do universal human rights begin?
In small places, close to home.
Eleanor Roosevelt, in Baird (1999, 75)

Domicide matters. It is a normal, everyday occurrence. At least 30 million people are currently suffering its ravages. The process of domicide and the effects on its victims are serious phenomena, and, as such, they deserve recognition in just the same way that both genocide and environmental concerns have received worldwide attention. As the world's leading expert on contemporary slavery states: "When people lose control of where they live and work ... they have lost fundamental human rights" (Bales 1999, 159).

We may well ask: Why is the destruction of home not prevented? This question leads to others. Why is someone not responsible for ensuring that the important personal values inherent in home are reflected in decision-making processes? Why is it not possible to find the convincing argument, to gather the necessary public protest, that would establish that the public interest lies in the preservation of home rather than its destruction? And, of greatest importance in an age of endless development, if destruction of home has to occur, why aren't the victims involved in a more participatory way so that they understand what is going on, can make serious changes to the plans, and thus move toward acceptance with much less trauma?

The short answer to the above questions is that we live in a world that an outside observer might well consider insane, a world controlled by power elites who operate according to a "command-and-control" agenda. When suffering is caused, the "react-and-cure" approach is wheeled in. Development and planning are much like allopathic medicine, designed not to prevent injury, but to wait until it has occurred and then try to cure it. These approaches are as foolish as they sound, but they are normal, which confirms Gruen's (1992) diagnosis of *The Insanity of Normality*.

In trying to reshape normality, which is the goal of this chapter, it must not be imagined that we are battling against change. Long before Heraclitus realized that "all is flux" around 500 BC, the Sumerian Utnapishtim declared, five thousand years ago: "There is no permanence. Do we build a house to stand for ever, do we seal a contract to hold for all time? From the days of old, there is no permanence" (Sandars 1960, 104). The point of this book is that houses indeed do not endure, but the decisions that lead to their destruction should come from the dwellers and owners, rather than be imposed upon them by external arbitrary force.

The issue, then, is that change should be guided not merely by far-off power elites for the common good, but by people living locally so that, at the very least, little harm befalls them. The range of what might be done is enormous, beginning with the realization that some projects, mostly those of everyday domicide, do indeed have merit, will enhance the common good, and may even improve the lot of those to be displaced. Most projects with the potential for domicide, however, will require larger or smaller modifications to minimize impacts on the victims. Others, especially some super-dams, simply should not proceed.

At this point, we will develop some principles for development projects that would protect those "in the way" from suffering domicide. The United Nations Declaration of Human Rights does not provide such principles, except rather vaguely in Articles 9 (No one shall be subjected to arbitrary ... exile), 12 (No one shall be subjected to arbitrary interference with his ... home), and 17 (No one shall be arbitrarily deprived of his property). These rights stand or fall on the meaning of the word "arbitrarily," which is easily circumvented by those committed to domicidal projects, who point out that all their actions are legal and have followed due process. The United Nations Conferences on Human Settlements do not specifically address domicide either, being largely concerned with "shelter." They do, however, call for: recognition that a human settlement is more than a grouping of people, shelter, and workplaces; the basic human right of people to participate in shaping the policies that affect their lives; and for high priority to be given to the rehabilitation of expelled and homeless people who have been displaced by natural and human-induced catastrophes.

Recognizing domicide as a catastrophe brought about deliberately by human agency, we would prefer that elites and their planning functionaries learn to act humanely, rather than ruthlessly. Justice Thomas Berger asked certain fundamental questions in relation to a proposed pipeline in Canada's Mackenzie valley: "What is the purpose of the project? In whose interests is it being undertaken? How should the economic gains be shared? Can the negative impacts be ameliorated or mitigated? We must even ask ... is the project the best way to do it?" (Berger, 1986, 181).

We might reformulate these questions as a set of fundamental principles:

- The first question to be asked is whether the project is really necessary.
- Incremental decisions, which ultimately force project approval, will not be permitted.
- Before any decision to implement a project is made, all those likely to be affected by it in terms of domicide should be involved as partners in the decision-making process.
- Before any project is implemented, detailed plans for relocation and compensation should be drawn up in partnership with those to be relocated.
- After implementation, the welfare of the relocated should be assisted and monitored at least into the second generation.

Such basic principles would revolutionize the way future projects are carried out, and domicide would disappear along with victimhood. Many projects would simply not be built.

The ultimate aim is that when development occurs, there should be no victims; in other words, we are looking for ways to prevent domicide without bringing development to a halt. In terms of implementation techniques, three chief approaches suggest themselves: the improvement of existing mitigation techniques; the adoption of alternative techniques; and the development of resistance on the part of the plannees. The latter, clearly, is most important in cases in which foolish and destructive projects must be utterly prevented.

EXISTING PLANNING TECHNIQUES REVISITED

We cannot know why the world suffers. But we can know how the world decides that suffering shall come to some persons and not to others. While the world permits sufferers to be chosen, something beyond their agony is earned, something even beyond the satisfaction of the world's needs and desires. For it is in the choosing that enduring societies preserve or destroy those values that suffering and necessity expose. In this way, societies are defined, for it is by the values that are foregone no less than by those that are preserved at tremendous cost that we know a society's character.

Calabresi and Bobbitt (1978, 1)

The first step is to decide whether domicide-producing development should occur at all. The second is to find out how far the visions of the elites differ from those of the public, and especially from those deemed to be "in the way." In this regard, there are a number of improvements that could be made to existing decision-making processes, particularly in public meetings and hearings and in social impact assessment procedures (which were discussed in chapter 6).

Public Meetings and Hearings

Public meetings and hearings are two quite different processes: the first is for gathering information; the second is the quasi-judicial stage prior to a final decision. Nevertheless, there are some improvements that could be made that apply to both. Both should allow for a specific period for hearing all persons who will be directly affected and provide financial assistance to permit their attendance. If these persons cannot attend the meeting or hearing, an independent assessor should be charged with providing a record of their views. Participants should be provided with research time and financial aid, coupled with the opportunity for public education and comments. Adequate notice of meetings or hearings should be given, and every effort should be made to ensure that the actual event is conducted by an impartial board in a non-threatening, non-legalistic atmosphere. Participants should have easy access to all relevant information, and an information meeting summarizing the material gathered during the process of project development should be held for hearing participants just before any formal public meeting. In this way, those who are charged with the task of approval or non-approval would have one final chance to decide whether they have adequate evidence on which to move to public hearing. An early example of an excellent public inquiry process was the Mackenzie Valley Pipeline Inquiry, which went to thirty-five communities in northern Canada and heard evidence in seven different languages. The report, which recommended against the enormous Alaska–Canada–US pipeline, was significantly entitled *Northern Frontier, Northern Homeland* (Berger 1977).

Social Impact Assessment

Social impact assessment is the area of anticipatory planning that appears to require the greatest attention. Assisting decision makers who may have little difficulty arriving at personal opinions to arrive at a decision in terms of social values is the challenge (McAllister 1980, 9). Cost-benefit analysis has been used in this regard, but it relies primarily on monetary values and fails to represent equally all the people of the present as well as the interests of future generations.

To rectify this situation, it is suggested that multiple account analysis, and particularly social-cost accounting, be used as a method for valuation. There is currently insufficient knowledge about the value systems of community and neighbourhood (Korsching et al. 1980, 336) and, we would add, about the individual whose home is being destroyed, and how this system will condition adjustment at time of relocation. Korsching (1980, 336-7) recommends gaining knowledge about this value system through the use of a modified Delphi technique, (whereby the "experts" are, for example, members of the group who will lose their home) or by the use of a nominal process in which a group, individual by individual, creates a list of their concerns in communication with a group leader, and then priorizes these concerns. There are disadvantages to this process. Those affected by domicide might feel ill equipped to assess impacts when a solid background of scientific knowledge is involved or might find it difficult to participate in such a process in view of the time involved. Nevertheless, it is suggested that in the context of the Delphi system or in social-costing discussed below, using "experts" from among those affected should be essential to augment other input. Such "experts" are, of course, inperts (Porteous 1977).

Social costing recognizes the need to achieve better enunciation of intangible values: "By working with people, by informing them of the issues, and assisting them in evaluating the trade-offs associated with the problem at hand, there is increased willingness to participate in the surveys. As well, this approach escapes the need for market information and does not require people to pull a price out of their head, which is the premise of the contingent valuation methodology. Overall, this approach answers the right question; [that of] the loss to our society and our region for a given decision" (McDaniels 1993, 21). The current focus of much of the discussion on social costing is on environmental factors and methodology. Some persons champion the cause of contingent valuation, whereby willingness to pay for public goods is measured (Kahneman and Knetsch 1992, 57-70), while others favour multiple objective approaches whereby people are asked to consider one value relative to another (McDaniels 1993, 21). With the latter, valuation is the final phase of a five-stage approach in which the "objectives" that matter are clarified, alternatives are structured, the impacts of alternatives are analysed in terms of the objectives, trade-offs associated with the alternatives are clarified, and the alternatives are evaluated. There is an opportunity for participation by those who would be most affected at each of these stages. The importance of this type of valuation is the recognition of externalities and so the impact on individuals of loss of home, surrounding landscape, and community, including the impact on future generations can be added to the lengthy list of impacts (e.g., on resources, the environment, land use, health, regional economic structure, and aesthetics). While the subject of social-cost accounting in relation to social impact assessment would require much greater consideration, it is simply suggested here that this emerging field may provide some guidance to persons who must determine whether projects will proceed and under what conditions.

It is not expected that improvements on existing techniques will be sufficient to produce an equitable development situation. While the 1960s saw a wave of citizen participation in planning, the 1980s was a decade of sober re-evaluation of earlier expectations that had in many ways not been met. Johnson (1984, 214) submits that citizen participation in decision making, even at the local level, is hampered by the scope and complexity of planning tasks, the political forces arrayed on all sides of the question, and the difficult and obscure structures and procedures through which citizens must act. He concludes that "the notion that ordinary citizens can influence the decisions of government as individuals or in groups of their own choosing is a powerful democratic myth," but its promise is rarely achieved because of non-involvement, the concern of citizens only for the "here and now," and the failure to articulate a vision of desirable futures. Meanwhile, elite visions prevail, offering vast future benefits that rule out as irrelevant any complaints about massive housing destruction (Davies 1972, Johnson 1984, 217). Catanese (1984, 36–7) adds that communication between planners and plannees remains difficult, not only because of the often powerless inarticulacy of the latter, but because planners continue to provide them with information that is technical and quantitative even when their concerns are clearly social and qualitative. Further, he cautions that citizens in democracies such as the United States are "inherently distrustful of co-operation and sharing," so that secretive elites are the norm, while participatory planning fails because "most of the people do not participate in it until it hurts." By this time, of course, it is usually too late.

Simmie (1974) is even more pessimistic, arguing that planning is not an objective, rational, apolitical professional activity conducted in the public interest, as it purports to be, but it is, in fact, a political process whereby planners choose to serve certain visions and interests and deny others. Simmie argues that traditional planning has come to a dead end and must learn to incorporate quite different structures of power, asking the questions: What would happen if there were no plan? Can people plan themselves? These questions lead us, first, to alternative techniques in planning, and, second, to resistance.

ALTERNATIVE PLANNING TECHNIQUES

Democracy is inconceivable without organization ... Organization implies the tendency to oligarchy. As a result of organization, every party ... becomes divided into a minority of directors and a majority of directed.

Michels, in Johnson (1984, 215)

Michels's "iron law of oligarchy" pervades more than the political parties he studies. Nevertheless, the sheer volume of citizen participation after the 1960s has led to a re-evaluation of the public role by both elites and planners, so that alternative approaches have become ever more acceptable. In Tinder's (1980) terms, pre-emptive authority, which thinks and acts for others and closes off the criticisms of those who are ruled, is giving way slowly to dialogical authority, which inspires discourse between all involved parties on the goals and methods of action. By the 1990s, it was difficult to approve projects of many kinds without lengthy discussions involving stakeholders of every kind. It is interesting to note that in the 1990s, even business journals were beginning to ask questions about the social costs of business decisions (Ward 1993) while realizing that the internalizing of social costs would require an enormous change in mindset for those previously committed to the ideology of the free market. Though true for the Western world, of course, this process is only in its infancy in the remainder of the planet.

Planning Approaches. A variety of planning approaches can be used to bring the concerns of potential victims to the attention of the decision makers and the media and to facilitate their real involvement in the decision-making process. These techniques are most useful when development is likely to go ahead but requires modification on the basis of direct input from those likely to be impacted.

Planning. Planning with and for persons who must be relocated could gain much from several techniques, including: integration of the social and environmental impact process with the whole planning process; the use of public participation techniques such as advocacy planning; and the use of strategic planning with or without environmental dispute resolution techniques. In particular, the goal of such planning should be to minimize the uncertainty surrounding the time frame of relocation and maximize the involvement of those who will lose their homes. Lawrence (1992, 25, 26) suggests that integration of environmental impact assessment (presumably including social impact assessment) would permit data, research, and experience to be utilized fully in the planning process and would provide cost saving as well by avoiding two separate processes.

As with social impact assessment, and as illustrated in the work of Arnstein (1969), Fagence (1977), and many others, it is also crucial that the people most affected be involved in decisions about their future. To ensure this involvement, use of the nominal process to design implementation options is suggested, as is "fishbowl planning," through which brochures, public meetings, and workshops continually revise and develop alternatives (Korsching 1980, 336). The Charette process could also be used, in which citizens, planners, community representatives, and politicians work together within a specified time frame on a community design (Fagence 1977, 301ff.). These methods could be augmented by other public participation tools such as attitude surveys and other means of provision of information and gathering of feedback. In particular, it is suggested that the appointment of an advocate – ideally someone from within the community who is a constant presence relating the interests of those whose homes would be lost – would be a useful extension of the concept of advocacy planning. The benefits of having people participate in decision making are numerous. Marris (1974, 99) believes that people lose their irresponsibility, improve their attitudes to administrators, sacrifice apathy or dissent, yet still interpret their situations as those without power would do. While this list may be overly optimistic, it is hoped that the definition of the project might change because of such participation.

Strategic Change Management. Today, it is unusual to be part of a large organization that has not been subject to some sort of strategic change management exercise. For these writers, who have been involved with several such exercises, there appear to be a number of lessons to be learned that would be of benefit to those who must deal with the victims of domicide. While there is now an extensive literature on the subject of strategic planning (for example: Bardwick 1991, Mills 1991, Moss-Kanter 1983, Peters 1987) and an equally large body of critique (Mintzberg 1993), reference to the work of a practioner is useful for this discussion (Haines 2000).

Perhaps most important is the recognition of the "roller coaster of change" that persons involved in change must ride. During the period in which change occurs, there are a number of emotions, similar to those experienced in bereavement, that people can be expected to feel and actions that they are likely to take. Upon the first impact or knowledge that change will occur, the reaction may be one of shock, fight/flight, or mourning. This may be accompanied by numbness or disorientation, a search for what is lost, and nostalgia for what has been. This stage is followed by a period in which there may be feelings of rage, anxiety, guilt, shame, and depression. The focus is on the uncertainty of the future; in such situations people often appear perplexed. During the next period, the search for a new future begins, and gradually hope returns and new energy is found.

Building upon the knowledge that the above reactions are likely to occur, strategic planning sets out, through a series of activities, to directly engage the persons whose circumstances will be changed. These activities, which could be adapted for use by persons who must relocate to a new area, include: developing a practical vision of the future; analysing the obstacles to achieving the vision; determining the strategic actions required to achieve the vision while still taking into account the obstacles that exist; determining the tactical actions required to carry out the strategies of the group; and developing an implementation plan of "who, what, where, when, how" for each tactical action for the first implementation period. This step-by-step approach has the advantage of breaking the tasks into small enough chunks so that people can visualize success while allowing for plan revision at regular intervals. Overall, the success of strategic change is likely to rest on a number of factors, including: the commitment of the leadership; the development of a vision by group leaders, which is shared and supported by the group, and which is comprehensive and detailed, positive, and inspiring; the ability to undertake strategies in the future, but to understand them through the past; the recognition that people do not resist change but resist loss or the possibility of loss (and that it is necessary to permit time for mourning the loss of the current state); and clear and continual communication throughout the process.

Processes such as strategic change that depend on social learning by community groups have a number of advantages (White 1987, 162–3). They permit the gathering of essential expertise from the people who will be affected; they create a momentum for changing government organizations and for promoting learning; and they increase the community's capacity to contribute to the development and its capacity for effective action. There are, however, certain disadvantages, including: the problem that there is no possibility of "no change;" you are either in the process or you're out; the danger of achieving mere participation as opposed to involvement (the highest rungs on Arnstein's ladder); the possibility that certain elites will dominate a process; and, finally, the possibility that the process will be too complex for those who are already trying to cope with losing their homes.

Despite these difficulties, the use of strategic change processes appears to provide a number of strengths beyond those described above. These strengths arise from knowledge that a loss will occur. They engage the participants in working toward a self-defined positive future. Here, knowledge of personality traits, as discussed by Simpson-Housely and de Man (1987), would be most useful, for those having the best ability to deal with change would lead such a definition. Finally, they emphasize the importance of communication throughout the process and the setting of an implementation and monitoring schedule to ensure that what is planned will occur. Strategic change management should become part of the planning process and should include planners who have undertaken special training in this area. Use of strategic change management would have a number of strengths in dealing with the victims of domicide, including: recognition of the stages of change that people must go through from the troughs to the peaks of emotion; the use of visioning to plan for the future; and perhaps most importantly, the design of a monitored implementation strategy that should continue long after domicide has occurred.

Environmental Dispute Resolution Techniques. Finally, it is suggested that when determining the planning processes to be used, there are lessons to be learned from environmental dispute resolution techniques. These techniques commonly involve the following characteristics (Crowfoot and Wondolleck 1990, 19): voluntary participation by those involved in the process; face-to-face interaction among the participants; and consensus decisions on the process to be used and on any settlement that may emerge (sometimes with the help of a skilled neutral mediator). While time-consuming, environmental dispute resolution techniques are seen as providing greater power and influence in the decision-making process as well as greater access to decision makers. Individuals also gain new skills in negotiation, communication, active listening, group process, and coalition building, as well as a sense of empowerment. These benefits would be very useful to persons who must change their home and sometimes rebuild their livelihood following domicide.

Softening the Blow

Beyond involving potential domicide victims in the decision-making and planning processes, a second set of techniques would be valuable in softening the blow. Three possibilities are: grief training; victim impact statements; and people's histories.

Grief Training. Grief training is the first of these measures. While it does not necessarily conform to the series of stages set out by Kübler-Ross (1992, 23) and others, Marris (1974, 23–33) portrays bereavement as "the irretrievable loss of the familiar" and describes the various symptoms of persons experiencing bereavement as including restlessness, exhaustion, illness, a withdrawing from people, a feeling of aimlessness, guilt, hostility, an inability to surrender the past, and a clinging to possessions. Three general reactions include a desire to escape from everything connected with bereavement, a worsening of physical and mental health, and a refusal to mentally surrender the dead. Marris further suggests that bereavement occurs, not because of the loss of others, but because of the loss of one's identity, and that intensity of grief is closely related to the intensity of involvement. It is important to accommodate all of these factors when trying to assist persons who must lose their homes. There is a significant link between home and identity, and this link is almost always stronger when there is greater involvement in the creation of home – for example, when a person has worked the land.

Another perspective is provided by Cohen and Ahearn (1980, 73), who suggest that when persons lose their home to a natural disaster, the following characteristics may occur: loss of self-confidence and inability to deal with the new situation in which they find themselves; a sense that everything in the future will result in failure and, therefore, feelings of guilt and shame; feeling "singled out" by the disaster even though they can see others in a similar situation; feelings of resentment when help is not forthcoming from those they expect it from; increased dependence on others; and loss of faith in group values. Where the victims of domicide differ from the victims of natural disaster is in their additional feelings of directed anger. Their fate is not caused by an "act of God" but by human action; therefore, someone is to blame.

Marris believes that survival in any situation depends on the ability to predict events. Such prediction is most successful when we are placed within a context that we understand, within the structure that each of us has learned from past experience. Simpson-Housely and de Man (1987, 3) corroborate this finding in their studies of natural disaster victims. They confirm that knowledge of personality traits and response to natural hazards enhances understanding of human appraisal of the hazards concerned. In essence, "life will be unmanageable until the continuity of meaning can be restored" (Marris 1974,

41) and until we learn that what has been lost can still give meaning to existence. On the other hand, Aiken (1985, 245) argues the need for a substitute for the loss in the form of: a reliance on religious or philosophical beliefs that stress the future; intensification of old social relationships or forming new ones; or becoming actively involved with other activities. Either way, a successful return to normality is the desired end: "Like death, the moment of transition is abrupt: the household wakes one morning in familiar surroundings and by nightfall is gone for ever. And as in grieving for the dead, all purposes and understanding inherent in those surroundings have to be retrieved and refashioned so that they still make sense of life elsewhere. If the new home is adaptable to their way of life, the adjustment is soon made ... The predictability of the new environment is not hard to learn, since it involves no radical revision of the past" (Marris 1974, 57).

What has been learned most recently in studies of dying and of grieving is the need to help people during this time through listening, counselling, and unconditional caring; through knowing when privacy should be respected and when intervention is necessary; through recognizing that conflict will occur; and through recognizing that both time and patience are required. All of these skills, utilized by trained professionals or lay counsellors such as social workers or hospice-trained therapists, would ease the transition when homes are lost and new circumstances accommodated.

Victim Impact Statements. The use of victim impact statements is borrowed from the criminal justice system and has been narrowly defined within that context to date. Since much of the literature on this subject relates to the procedures to be used and case studies of use, rather than the theory associated with use, this section is derived from discussions with Tim Roberts of Focus Consultants (Victoria), who is recognized as the British Columbia expert in this area.

From the nineteenth century, crimes were defined by the judicial process as an issue between the individual and the state, so that the victim was excluded from any part of the trial. Victim impact statements were developed recently to permit the person who had been most affected to have some say in the sentencing of the offender, while the state retains the right of judgement based on facts. Two benefits accrue: first, the statements sensitize the criminal justice system to the essence of the crime; and second, the victim may be helped by the therapeutic aspect of having his or her story officially recognized. The use of these statements is now leading to victim–offender mediation based on a tradition of restorative justice, which believes that relationships must be mended.

In the context of this discussion, victim impact statements would prolong the compensation process, a fact that should not detract from their use but that should be planned for initially. Beyond the usefulness of victim impact statements to assist in the establishment of mitigation when people lose their homes, there is a problematic theoretical perspective. The concept of victim–offender mediation allows a closing of the circle. However, in the case of the victim of domicide trying to speak to either government or a large corporation, there would always be the question of whether there was anybody "at home" at these levels to close the circle and mend the emotional rents of domicide.

People's Histories.   The creation of people's histories, as undertaken in the History Workshop movement and in oral and family history, is a technique that could be used effectively when it becomes certain that people will lose their homes. Life history has been defined as "the account of a life, completed or ongoing, the use of which can present an individual's evaluations of experience and give the context of experience" (Eyles and Smith 1988, 10). Plummer (1983, 13–38) identifies a number of different types of life histories. Of these, however, the most successful are those whereby people are able to tell their own story in their own way. As the Yorkshire Art Circus asserts: "Everyone has a story to tell. We find ways of helping them tell it" (Griffiths et al. 1988).

Porteous (1989, 232–3) has noted how ordinary people have seldom created autobiographies. Indeed, stories of domicide and the erasure of the past are most often incorporated into fiction, following the romantic "impulse to set free tongues tied or silenced by oblivion or oppression" (Hughes 1988, 94). This lack has been rectified to some degree with the new emphasis on the recognition of everyday life in the academic fields of history (particularly labour history), geography, and planning. Family histories contain an evocation of place as well as the motivations of individuals. In this way, they provide a valuable record of places that are to be destroyed and become a source of memory. Cragg (1982, 48), in undertaking family history, acknowledges that "nostalgia has become a key to determining what I valued in the landscape and what responses prompted my emotional and imaginative growth." Ebaugh (1988, 31–4) describes this process as permitting the experiences and definitions held by a person, group, or organization to be interpreted by that person, group, or organization. In so doing, it is possible to elicit the different ways in which each person interprets an experience and therefore reacts in a specific way. In gathering this information, the sequence of events, social context of the events, interpretation by the individual, and reasons for individual reactions are all important.

It is also important to recognize the possible shortcomings of this tool, including the possibility that a person's memory may fade or that present experience may influence past experience. However, these records are likely to be the only ones that will ever be available, and, once again, perception is reality. The telling of life histories also has therapeutic value as a palliative measure. Encouraging the victims of domicide to provide information about their homes enables special recognition of their circumstance and comfort that someone is listening. Such histories will quite simply prevent memoricide, the loss of individual feelings about a place that could otherwise disappear forever.

In their heartening *Listening For a Change*, the oral historians Slim and Thompson (1995) demonstrate the value of people's history based on case studies carried out among displaced people in the Sudan, Mozambique, and elsewhere during the 1980s. For the Dinka of the Sudan, the telling of stories and legends not only provided important information about how previous disasters had been weathered, but was also of great spiritual value to a community in crisis. The narratives of Mozambicans indicated that the displaced knew what they needed to make their survival easier and showed that they were the best placed to design their own relief programs. The authors contend that community voices need to be listened to very early in the displacement process, so that particular needs and priorities can be described. Thus, instead of providing the same package of support to every group, "more appropriately targeted interventions should be developed which build on a community's strengths and minimise their particular vulnerabilities" (Slim and Thompson 1995, 32).

Participatory Action Research

Finally, both planning approaches and strategies for softening the blow can be brought together in the process known as participatory action research, a complex approach that may involve elites, planners, citizens, and academics.

From the academic viewpoint, participatory action research stems from two movements: the recent turn toward qualitative research and the realization that academics can play a role in redressing the inequities of the current balance of power between elites and the plannees, who so easily become victims. Qualitative research came into its own in the 1980s in a variety of disciplines, both pure and applied (Gold and Burgess 1982, Lincoln and Guba 1985, Walker

1985, Eyles and Smith 1988). Practically, this has brought to the fore the fact that most people live their lives according to intangibles such as love or loyalty, rather than operating in a strictly quantitative economic and political nexus. Elites, scientists, and planners knew this intuitively, of course, but did not wish to acknowledge such concepts. As Wikström (1994) points out, his qualitative research on relocations in Sweden has made the lives of planners and designers more complicated. When put into action, with the explicit goal of effecting change, qualitative research becomes more complicated still, rather like life.

Action research is associated with the work of the psychologist Kurt Lewin in the United States in the 1940s. Lewin's (1946) emphases were four: the research should help solve social problems; it is an iterative process, requiring many repetitions to become efficacious; scientists require specialized training to translate data into practice; and intercultural communication between planners and clients must be facilitated. The essential feature of the approach was the simultaneity of research and praxis. The concept was refined by Sanford (1970), who defined action research as problem-centred research that bridges the gap between theory and practice. For Sanford, its three interrelated objectives are: to improve the situation of the participants; to advance knowledge; and to refine its own theory and praxis. Sommer (1997) pays particular attention to the need for action research training, facilitation of information flow between participants, and adequate post-research evaluation or follow-up.

In a parallel evolution, coming chiefly from sociology and organizational behaviour research, Whyte (1991) defines participatory action research (PAR) as a partnership between researchers and citizens. It explicitly rejects research models that identify scientific progress with survey research and quantitative modelling, since the subjects of such studies have no opportunity to check the "facts" as the researcher sees them. In contrast, PAR requires partnership between academic and citizen, so that the ownership of the research becomes more widely shared than in the conventional model. Just as findings are shared, so is theory, and work with the Basque co-operative workers at Mondragon conclusively shows that clients with a stake in the truth are quite capable of sophisticated philosophical and methodological discussion. Significantly, Scandinavians are far ahead of North Americans in acceptance of such methods, mirroring their commitment to worker participation in industry and their rejection of the hidebound North American command-and-control approach. In an appraisal of PAR, Wisner et al. (1991) document its value in fields as diverse as disease-control, social welfare, farming, resource management, and house design programs. Of particular relevance to domicide is the world-renowned work of Turner (1984) in the field of self-built housing by Third World squatters.

Directly relevant to domicide is the PAR experience at North Bonneville in Washington State. When the US Army Corps of Engineers decided in 1971 to site their new hydroelectric powerhouse on the site of the town of North Bonneville, they told the 470 residents that they would have to relocate from the area (Comstock and Fox 1982). Because, according to the researchers, they were politically conservative and fiercely patriotic, citizens did not try to stop the project. Nevertheless, the highly independent but socially cohesive community decided that they wished to relocate as a community to retain both their social bonds and their physical relationship with the Columbia River Gorge. Consequently, the community enlisted the assistance of the Bureau of Community Development of the University of Washington to form the North Bonneville Life Effort (NOBLE), which produced an initial survey showing that residents felt a strong identification with their town and the surrounding landscape. Subsequently, assistance came from faculty and students of Evergreen State College, who used a Latin American model of social change to empower citizens through PAR.

Students lived in the town and shared the problems and goals of its citizens. As the groups interacted, residents came to realize the huge discrepancy between their knowledge of who they were and the very different perspective of the Corps of Engineers (hooked on technical efficiency) and the politicians who wrote the relocation laws (concerned mostly with compensation and efficiency). While residents came to define their community as a collective "home" – a complex network of social, natural, and spiritual relationships – they discovered that planning elites saw their town merely as a quantifiable number of individuals and physical artifacts. They also discovered that their goals, intangible and social, were quite different from those of the government, which was to build the powerhouse as quickly as possible with the least uproar.

These realizations helped citizens to oppose the corps' original plans, demand access to withheld information, and produce a feasible alternative plan. In a series of community-wide workshops that involved all residents in decisions, the community took upon itself the tasks of finding a new site and designing the new town. Despite the efforts of the Corps of Engineers to dissolve, ignore, and subvert the relocation contract, the town turned to the courts to force the corps to fulfill its contractual obligations. The whole process took from 1972 to 1976, when the new city of North Bonneville was officially dedicated.

This experience demonstrates that domicide can be avoided. The citizens of North Bonneville saw their old town demolished, but were rewarded with a nearby new city, designed by themselves. Although student researchers taught them some skills and helped organize research, citizens were emancipated by the process and empowered to make their own decisions. The ability to challenge the US Army and win was a powerful source of self-esteem, valuable in dealing with the private businesses contracted to build the new settlement. As community power grew, student facilitators withdrew, their work done.

The Bonneville experience demonstrates the potential for PAR to initiate a revolutionary process whereby citizens are empowered to research and critique the values inherent in a plan, generate an alternative scenario, and bring it to fruition. It is perhaps ironic that these highly conservative citizen activists pursued a course of action with roots in Mao Zedong's revolutionary thought, Gandhian social action, Latin American liberation theology, and Tanzanian self-reliance theories (Wisner et al. 1991, 273).

Wherever possible, of course, the most beneficial tool to assist people who are to lose their homes is a direct share in the benefits of project construction, as is now occuring in BC's Columbia River Basin. Yet, these benefits may take many years to realize, are often difficult to estimate, or may never materialize at all. The insights of strategic change management suggest that people do not resist change so much as they resist loss. Thus, the emphasis of strategic change management is on envisioning a desirable future, recognizing the threats to, and positive means of, reaching that future, and establishing a firm program for implementation. All this fits very well with the battery of techniques (Wisner et al. 1991, 275) used in participatory action research.

No single measure discussed above provides a sufficient answer, and the use of some would be problematic when large numbers of people have to relocate. Nevertheless, loss of home should be addressed at the outset of any project, and a suitable set of approaches, always involving some form of PAR, should be decided upon. Failure to do this will increasingly evoke resistance.

RESISTANCE: DO NOT GO GENTLE

> Now be quite calm and good, obey the laws,
> Remember your low station, do not fight
> Against the goad, because, you know, it pricks
> Whenever the uncleanly demos kicks ...

> It's fearfully illogical in you
> To fight with economic force and fate.
> Moreover, I have got the upper hand
> And mean to keep it, do you understand?
>
> Hilaire Belloc (1912, 36)

Belloc's lines remind us that, above all, victims of domicide, like those of colonialism, downsizing, or the myriad shocks that affect ordinary people, are ultimately dealing with power. Opposed, power fights back.

As one NGO report asserts: "The quality that most defines humankind throughout history is this quality of resistance, and the capacity to struggle beyond the limits of what is human and what is possible, against all odds" (Inter Pares 1994, 1). Edward Abbey (1984, 1988) is eloquent about resistance, while John Pilger (1998) listens to what political and business leaders say, watches what they do, and then contrasts the two. In John Sweeney's words, "Journalism – the real thing, not crap about the sex lives of the Royal Family – means poking people of power with a stick and seeing what happens next" (1998, 14). For opposition is essential, and resistance is necessary whenever the elite try to ride roughshod over people "in the way" of their projects. The success of North Bonneville could not have happened without the will to resist the US Army Corps of Engineers. Without resistance, people become wounded human material who "only know how to think and behave as victims" (Makiya 1994, 225).

Resistance to power arises with what Latin American theorists call "conscientization" – the bringing to our consciousness that, one way or another, we are oppressed by authority. In terms of domicide, people must come to realize that homes exist at the whim of corporate, bureaucratic, and political power elites, and that, should these groups so decide, communities can be eliminated and individuals forced to relocate. In particular, the common good rhetoric must be questioned, for in Segal's words: "Law and order, or national survival, seems less and less adequate an excuse for the ravages of the social and natural environment, the subjugation of the person to profit ... More and more ... citizens must ... ask for whose good the state really operates" (1973, 192). As Gold and Burgess (1982, 3) remark: "In many instances, the true nature of the 'public interest' remains obscure, a fanciful conception occasionally invoked rather than a goal to be actively sought ... it is not necessarily promoted by the imposition of schemes that ignore the needs of local communities in order to serve some notional concept of the common good of society at large." Most of all, we must resist the claims of elites and planners that they already know what people want; a vast amount of research demonstrates that such is rarely the case.

Resistance refers to actions imbued with the intent to "challenge, change, or retain particular circumstances relating to societal relations, processes, and/or institutions. These circumstances may involve material, symbolic, or psychological domination, exploitation, and subjection" (Routledge 1997, 361). The concept of resistance is heavily theorized, the points of view ranging from Gramscian theories of hegemony through critical social theory to postmodernism (Routledge 1997, Dear and Flusty 1998, Myers 1998). Gramsci (1971), in particular, argued that the dominant elite's position and function in the world of production, together with its moral and intellectual leadership, generated a prestige that led to the spontaneous popular consent that lies at the heart of hegemonic power. It is our loss of respect for elites, our withdrawal of consent, that permits us to resist hegemony. It is the sheer obviousness of oppression, clear to all who have suffered domicide, been downsized, or otherwise had their lives forcibly changed, that leads us to seek equity, fairness, and justice.

The roots of modern resistance lie in the social upheaval and brutalization that accompanied the industrial revolution, with yet deeper roots in those eighteenth-century Enlightenment philosophers who began to realize that social institutions are not God-given but historical constructs subject to subversion and change. In the early nineteenth century, those who took existing power relations for granted argued for social reform. More radical approaches – notably those of Utopianism, social anarchism, and historical materialism – argued for oppositional social mobilization (Friedmann 1987) based on social emancipation. From Utopianism came the vision of an alternative future, from anarchism came the idea that society could be organized on the lines of peaceful co-operation rather than aggressive competition, and from historical materialism came the realization that a historical change can be looked upon as a class struggle between the dominant and the oppressed, the rich and the poor, the proponents of domicide and their victims.

In the twentieth century came the realization that although oppressive systems may be global, emancipatory struggle "is always particularized and local. It involves particular individuals and groups, in particular situations, facing particular problems" (Friedmann 1987, 301). In the words of the American philosopher John Dewey: "Democracy must begin at home, and its home is the neighborly community" (1946, 213). It is this truth that lies behind the discussion of resistance, not only in general terms, but also in terms of the lessons we can learn from individual cases such as the Chipko and Balipal movements in

India (Routledge 1993), the resistance to road building in Britain (Routledge 1997), and the use of music to resist state propaganda in Singapore (Kong 1995).

Resistance is an idea whose time has come. Epochal moments in the history of resistance include Thoreau's "civil disobedience," Gandhi's *satyagraha* (soul force), the flocking of anti-Fascists from many nations to the Republican cause in the Spanish Civil War, the "Flower Power" rebellion, which helped bring an end to the Vietnam War, the socially dynamic religious philosophy of Martin Luther King Jr., the flocking of idealistic youth to pick cotton in Nicaragua after the 1979 Sandinista revolution, and in the face of US aggression, the women of Greenham Common, UK, who helped undermine the nuclear military agenda, the millions of Filipinos who non-violently overthrew the Marcos dictatorship in 1986, the intifada in Palestine, which forced the Israelis at least to consider the possibility of Palestinian self-rule, and more recent demonstrations against the World Bank, International Monetary Fund, and other economic globalization agencies.

The year 1989 was the year when resistance came into its own. The intifada was changing Israeli politics, the campaign to free Mandela and destroy apartheid was reaching its climax, and Berliners demolished the hated wall in a symbolic gesture that heralded a completely new global geopolitics. But the seminal resistance image of the twentieth century will surely be the unknown student who stood firm before a line of tanks during the democracy protest in Tiananmen Square, Beijing. Previous generations have perversely revered the "unknown soldier," sent to his death for the purposes of power elites. The twenty-first century will surely turn their eyes to the "unknown student," sure of what he wants, convinced that it is morally good, and willing to fight the massive power of government to achieve it.

Recent surveys of active non-violence (Deats 1997), indigenous resistance movements (Pascal 1997), celebrations of resistance (*New Internationalist* May 1996), and compendia of grassroots actions (Ekins 1992) confirm that resistance movements are providing a seedbed for social ferment and revolutionary change throughout the planet. Though sometimes led by external activists, they are just as likely to come up with a *Local Hero* (Benedictus 1983). They forge coalitions and links with local, regional, national, and international organizations while maintaining their local integrity. They learn to use the media to convey their messages of resistance to both the broader public and to the authorities. Their techniques are legion, from sit-in occupations of public property and mass marches to community quilt-making and parish map making. Street theatre is a common device locally; Internet connections work globally. Bioregionalists everywhere encourage the mapping of home (Aberley 1993). Women are often at the forefront, most notably in the Third World (Shiva 1994).

The civil strength of the resistance movement is inspiring. Sem Terra, a Brazilian organization of the landless, has 220,000 members and enjoys the support of 90 per cent of the nation's population (Vidal 1997, 8). The Philippines has 18,000 registered NGOs, two-thirds of which are people's co-operatives. The state of Tamil Nadu, India, has 25,000 registered grassroots organizations. Women activists worldwide presented a petition to the UN World Conference on Human Rights in 1993, which carried 500,000 signatures from 124 countries. Major UN and international conferences are now routinely mirrored by parallel alternative conferences with resolutions, relatively free of diplomacy, that often make better sense. Trade union membership, though suffering setbacks in Britain and the United States, has surged in newly democratic countries in Eastern Europe and Africa. In cities, in which most of us now live, radical planners see the rise of civil society in a global age as heralding new opportunities for bottom-up planning.

Two points must be made clear. First, resistance is almost wholly non-violent. The stone throwing of the intifada is not common elsewhere. Though Earth First! activists use monkeywrenching ecotage tactics (Porteous 1996) against machinery and property, they do not harm people. The moral strength of resistance involves laying down fragile bodies before bulldozers. Violence sometimes occurs, but it is almost always carried out by the authorities against non-violent demonstrators. Second, resistance does not normally indulge in nationalist public interest or common good rhetoric. It is more likely to speak of the good of particular groups or particular localities, or at the other end of the scale, the common good of the global environment.

Resistance to Domicide

The worldwide resistance to government and business oppression has a considerable part to play in resistance to domicide. From a wide array of examples, we choose to consider resistance to domicide caused by military activities, Third World "development," the redevelopment of Western city regions, and dam building across the globe.

Military activities in the Pacific have been major causes of domicide. Having lost its bases in Vietnam after its defeat there in 1975, the United States military fell back on bases in the Philippines, from which they were ousted in turn in the early 1990s after a long grassroots campaign and the co-operation of a volcano that spewed deep mud over Clark Air Force Base. Requiring a fallback position in the 1970s, the US chose the tiny nation of Palau (Belau), a trust territory that the United States was mandated by the UN to bring to independence.

Palau, a nation of about 17,000 people east of the Philippines, had strongly resisted signing a Compact of Free Association with the US, and had instead written the world's first constitution declaring itself a nuclear-free territory. This constitution required a 75 per cent majority vote to ratify an amendment to permit US bases in Palau. From 1983, the United States forced the nation's people to hold referendum after referendum in its pursuit of the right to create military bases. Anti-nuclear activists and one Palauan president were murdered by pro-base Palauans, and another president committed suicide. Washington denied much-needed funds and delayed Palauan independence, effectively blackmailing the Palauan people. All eight pro-military referendums were rejected until 1992, when an amendment was passed requiring only a simple majority for constitutional change. Thirteen months later, the ninth plebiscite since 1983 was finally passed by less than half of Palau's eligible voters (Thomas 1995, 173). The United States now has the right to conduct military exercises, including nuclear ones, on about one-third of the nation's territory. This will have a profound effect on national self-esteem and identity, and it is being fought by the Palau women's movement, *Otil a Belaud*, and many others who "do not want our land to be someone else's 'contingency plan'" (Denoon 1997, 335).

A more successful case, however, occurred with a peasant movement against the siting of a missile-testing base at Balipal, Orissa state, in India. A highly stratified society with no history of peasant insurrection, Balipal felt sufficiently threatened to transcend barriers of caste, class, gender, age, and political affiliation. The movement made heavy use of religious symbolism, including the invocation of Kali as both warrior goddess and earth mother. Women were prominent members of the movement, which expressed itself in an ideology of *bheeta maati* (our soil, our earth, our land, our home) (Routledge 1993).

Third World movements constantly hark back in this way to traditional stories of resistance and deeply embedded cultural practices of self-help and collective co-operation. Routledge (1993, 1997) gives many examples of this from every continent. Threatened with the wholesale burning of homes, crops, and rubber trees by ranchers and speculators, the rubber tappers of the Amazon joined forces with indigenous peoples to protect the latter's homelands and the former's homes and livelihoods. The Orang Ulu peoples of Sarawak in Malaysia, who look on the forests as their traditional homeland, blockaded logging roads against the logging corporations, one of which is owned by the Malaysian minister for Environment and Tourism. Ekins (1992) provides another array of resistance attempts worldwide, some of which were successful. The Sarvodaya Shramadana Movement is active in 8,000 of Sri Lanka's 23,000 villages, and it has benefited millions of people by promoting grassroots development and resisting large-scale changes proposed by outsiders. The West can clearly learn from the Third World; the earlier success of the Bangladeshi small-loan Grameen Bank has been copied by Shorebank – a local initiative in one of the poorer parts of Chicago – which took over the South Shore Bank of Chicago and dedicated it to a revival of the neighbourhood.

The formerly passive inhabitants of Western cities have also been moved to take power into their own hands. In the 1960s, protest against redevelopment was common. Lupo et al. (1971) explain in detail how the threat of inner-city expressways and the devouring of East Boston by Logan Airport brought previously separate sectors of the population – citizens, politicians, and academics; Jews, Irish, and Blacks – together on the Greater Boston Committee on the Transportation Crisis. Transcending their educational, class, and ethnic differences, these groups co-operated in community organization by researching, lobbying, and planning to save their neighbourhoods from destruction. Before this outburst of the late 1960s, the decisions to build city expressways in Massachusetts were taken by autonomous state and city agencies unresponsive to the public and replete with common good rhetoric. Afterward, public involvement could not be ignored. Similar movements sprang up across North America, and by the late 1970s, neighbourhood-destroying road building was grinding to a halt everywhere. Goodman's (1972) *After the Planners* recounted similar success in the United States and Britain, and it looked forward to an era when citizens could be free from forced relocation and plan for their own communities.

More surprisingly, perhaps, there was considerable opposition to Soviet centralizing and modernizing tendencies, particularly in the outlying republics of the USSR. Uzbeks, for example, have traditionally lived in the *mahalla* (a distinct and identifiable social neighbourhood, district, or urban ward). Each *mahalla* is its own world, with its own rules. Uzbeks cannot imagine a person without a *mahalla*; such a person would be *mahalsiz* ("groundless, ill-founded, out of place; insignificant") (Malcolmson 1994, 179, 193). The *mahallalar* stood for decades as bastions against Russian imperialism – from land tenure laws through intermarriage to resistance to being decanted into highrise towers. The Uzbek word for this way of life, this intense identity-supporting and life-centring localism, is *mahallaçilik* (the state of "being-in-a-*mahallaness*").

Third World resistance has also been active in urban areas; the case of the squatter settlement of Sanjay Ghandi Nagar (SGN) in central Bombay is one of many. Brought to Bombay to labour on the high-rise blocks that create the Marine Drive skyline, the poor people of SGN were permitted to squat on a worthless piece of land reclaimed from the sea (Seabrook 1997). Expected to move on in 1979, they stayed and took up a variety of central-city occupations; many were servants in the well-to-do district of Colaba, next to their settlement. When the rich of Colaba demanded that the servants' unsightly homes be removed, the Bombay government destroyed the colony in 1989, but having nowhere else to go, the SGN squatters rebuilt it. The same sequence occurred repeatedly thereafter. With much of the Third World's urban poor living in slum or squatter conditions, often on land that is prime for redevelopment, the potential for government violence and citizen resistance is enormous (Jhabvala 1981, Mistry 1995).

Resistance at dam sites has gained far more attention worldwide than any other terrain of protest. Moreover, the spectacular nature of super-dam projects has captured international media attention and enabled strong alliances to be created between grassroots resistance and both national and international resistance organizations. Three super-dam projects are worth detailed consideration.

Because of the nature of Chinese governance, the Three Gorges project is most insulated from both grassroots and international protest actions. The danger of Chinese resistance is seen in the arrest and imprisonment of Dai Qing and the pulping of her book *Yangtse! Yangtse!*. Probe International, a Canadian NGO, has been at the forefront here, publishing *Odious Debts* (Adams 1991), an attack on World Bank funding for environmentally ruinous projects worldwide; a book on the Three Gorges project (Barber and Ryder, 1990); and an English translation of Dai Qing (1994). To expose the role of Canadians in the cases of both Three Gorges and James Bay in Quebec, Probe International and the Quebec Cree jointly prosecuted Hydro-Québec, BC Hydro, the Canadian, Chinese, and Quebec governments, and three private Canadian engineering firms before the International Water Tribunal in Amsterdam in 1992. The Tribunal ruled forcefully in favour, calling for a halt to both projects. Pressure continues on both environmental and social grounds, the probability of relocation success being highly unlikely in a command-and-control situation in a culture of poverty (Dai Qing 1994, 226; Jun 1994 259–60).

In the case of James Bay (Pearce 1991), the James Bay I project was built by Hydro-Québec in the 1970s. As completed, it comprises nine dams, five reservoirs, and 206 dykes, has diverted four rivers, cost $20 billion, and flooded over 30,000 square km of the traditional hunting territory of the Cree. The project involved several relocations, including the creation of the new town of Chisasibi on the La Grande River. The relocation of the Cree to this "piece of big-city suburb airlifted to the North" (Dwyer 1992, 36) was disastrous. "After the move," says Chief Violet Pachanos, "there was a great rise in alcoholism, drug use, family violence, and family breakups." The hunting life is no longer viable, and the promised jobs have not appeared. The Cree soon realized that they had been taken advantage of chiefly by Robert Bourassa, Quebec's late premier, who saw the massive export of electrical energy to the United States as the basis for greater Quebec independence from Canada.

In consequence, the Cree joined forces with other resistance agencies to oppose the James Bay II project, which, beginning with the Great Whale River phase, was planned to divert a total of twenty of Hudson's Bay rivers and flood 23,000 square km, affecting in some way a total land area of 350,000 square km. As in multi-ethnic Boston in the 1960s, such was this threat to traditional homelands that even the rival Inuit and Cree began to work together in the 1980s. This multiracial coalition made effective use of the national and international media and convinced the state and city of New York to refuse to buy Quebec's electricity. Resistance proved successful, and in 1994, the Great Whale River project was cancelled (Séguin 1994, 1–2).

An equally interesting and heartening case of resistance to super-dams is the story of protest against the Narmada River Project (NRP) in India. Clearance for building the two super-dams was given by 1988, but already those likely to have their homes flooded were questioning the project's plans for resettlement, compensation, and environmental protection. When it became clear that their conditions would not be adequately met, a wide variety of opposition groups, many led by women, made common cause and created one of the most powerful social movements ever to emerge in India's half-century of independence. Relying, like the Chipko and Balipal movements, on local and national traditions of resistance, religion, and the security of the hearth, mass demonstrations, non-co-operation, media projects, and international alliances were launched.

One early demonstration in 1989 had 60,000 participants who protested the coming submergence of the town of Harsud. Thousands of the demonstrators pledged never to move from their homes. These direct actions reached a large and increasingly sensitized international audience. At this international level, an extraordinary meeting occurred in 1992 when the World Bank was requested by Canada, the United States, Japan, Germany, Australia, and Norway to suspend loans to the Sardar Sarovar and Narmada Sagar projects of the NRP. This protest ultimately met with some success for, after lending $280 million, the World Bank pulled out of this project following the launch of an international campaign to cut off the bank's source of funding (Probe International, 1993b).

The World Bank made this decision on the twenty-second day of a hunger strike by a number of people, including a woman doctoral student, Medha Patkar, who led a march of 3,000 people from the Narmada Valley to the dam site hoping to stop dam construction. These are some of the same people who had officially, and impersonally, been referred to throughout the project construction as "project affected persons (PAPs)" and whom Ekins better identifies as "oustees" (Ekins 1992, 91). Even with the World Bank's departure, the effects of this project have continued, for India has pressed on with construction despite a Supreme Court injunction of 1995. Probe International (1994) has reported deaths in the resettlement sites already developed. Appalling living conditions have caused fifty of the eighty families sent to one site to return to their original homes.

Sought by the police, hounded by *goonda* thugs, her telephone bugged, Medha Patkar has followed Gandhian principles in her organization of the resistance coalition known as Narmada Bachao Andolan. Winner of the 1991 Alternative Nobel Prize, she regularly refuses international money offerings to retain grassroots control. Much of the work involves education, conscientization, and empowerment of the tribal peoples most likely to suffer from the flooding. In a fine display of local identity and national pride, a powerful and disarming combination for resisters, one newly literate girl wrote on her slate: "My name is Sampat and my abode is Pepulchup. Long live India!" (Powell 1994, 19). The goal is to save such tribal villages from destruction and the consequent annihilation of tribal cultures by a government that has made promises it cannot possibly keep.

Pressure continues. After months of preparations 10,000 Indian villagers, led by women, took over the partly built NRP Maheshwar dam site in January 1998 (Vidal 1998). The dam, the first privatized hydroelectric power project in India, will submerge the homes of 2,200 families in sixty-one villages and destroy thousands of hectares of cotton and wheat land. Incensed that the dam will benefit areas downstream but not themselves, villagers demanded an end to all construction and a project review with meaningful public participation. When the chief minister of Madhya Pradesh state officially halted construction on a temporary basis, villagers quickly reminded the press that "we have stopped the work, not him" (Stackhouse 1998, D4). The right to review was granted. Actions like this will continue until the NRP is either stopped, severely modified, or becomes an issue on which federal and state governments choose to violently overrule the wishes of major segments of the Indian population.

International resistance to dams has continued with the exposure as "a dam-building mafia" of several active dam building engineering, electricity, and political lobbies such as the US National Hydro Power Association and the International Commission on Large Dams. In 1991, outraged by projects such as the Itaipu dam that drowned the homes of 42,000 people and led to their being dumped thousands of kilometres away, Brazilian resistance groups coalesced as MAB (national movement of people affected by dams). Long-term research projects are demonstrating that while the lives of the relocated may improve initially, as with the Kariba dam on the Zambezi, poorer farmland and worsening social relations have resulted in a second generation distinctly worse off than their parents had been (*Economist* April 1997, 80).

Because of "a flood of protest" (*New Internationalist*, November 1995) and "a flood of fiascos" (*Economist* April 1997), and the work of organizations such as Probe International and the International Rivers Network, the building of super-dams in the developed North is grinding to a halt, while both Third World governments and lenders such as the World Bank are becoming distinctly cautious. Major projects have been cancelled in Thailand, Bangladesh, and Nepal (Angelo 1996). With the general slowdown in international funding, it is likely that the fashion for super-dams may give way to something more reasonable, such as Oxfam's "Give a Dam" campaign for small-scale locally engineered water catchment projects. Since the height of dam building in the 1960s, when the world built over 1,000 large dams a year, the figure had fallen to about 250 in the mid-1990s, and it may be less than one hundred by the early twenty-first century.

What is clear today is that no large dam can be constructed anywhere in the world without controversy, and that local residents facing domicide now have at their command a battery of tools of proven value in a grassroots-internationalist alliance of resistance against national governments and transnational corporations. Like the super-dam itself, this situation is a major achievement of the twentieth century, which may well ensure that the century's destruction of the homes of 60 million people for dams will not be repeated in the new millennium.

Indeed, staunch local resistance at a number of sites, coupled with international NGO pressure, led to the setting up of the World Commission on Dams in 1998. Commissioners were drawn, on the one side, from hydro power corporations, the International Commission on Large Dams, global engineering firms, and water resource ministries and, on the other, from Oxfam, the International Alliance of Indigenous Peoples from the Tropical Forest, the US Environmental Defense Fund, the Narmada Bachao Andolan of India, and also a number of academics from anthropology to engineering. The commission's chairman stated that its goal was to find solutions to the problems "that currently plague the relationship between those who plan, finance, and build large dams and those who oppose them" (Dorcey 1998, 26). The development effectiveness of dams will be considered, the "real costs" of dam building will be investigated, and guidelines for better decision making proposed. And, finally, "one of our aims ... should be to ensure that such projects in the next century also encompass broad public acceptability before any work begins."

In its final report, the commission noted a persistent failure to assess the negative social impacts of dams, and estimated that between 40 and 80 million people were displaced by dams during the twentieth century (World Commission on Dams 2000). The report went on to recommend 26 guidelines for the reviewing and approval of projects. Of major interest to this study was the concept that five core values should be basic to future decision making: efficiency, sustainability, equity, accountability, and participatory decision making. The latter is extremely welcome in the context of possible future domicide.

These, as yet, are mere promises. Broad public acceptability could still be translated as "the common good," thus ignoring those who actually suffer domicide. And capital is powerful; it may still be too early to agree with Stackhouse that we are already experiencing "the decline and fall of the big-dam era" (1998, D4).

We must also reiterate that dissidence and resistance are not for the faint-hearted, and they will probably work best in Third World situations in which people are generally not held hostage by their lifestyles, mortgages and pension plans. Numerous protesters have been killed, beaten, or imprisoned by governments during direct action. Even to write is dangerous in many countries. Abdul Rahman Munif, author of *Cities of Salt*, has had his books banned in several countries and, for his portrayal of Saudi–American collusion in domicide, has been stripped of his Saudi citizenship (Said 1994, 289). When facing overwhelming odds, however, resistance groups can always recall the stirring Hebrew story of David and Goliath, which brings us face to face with the wider context of domicide, namely, the power of globalizing capitalism.

RE(S)TRAINING CAPITALISM

Many ... see "the state" or "the planners" as their essential enemy, when it is quite evident that what the state is administering and the planners serving is an economic system which is capitalist in all its main intentions, procedures, and criteria.

Raymond Williams (1975, 353)

One of the saddest facts of archaeology is that the majority of the most ancient texts from the cradle of civilization, laboriously excavated from the sands of Mesopotamia, are no more than a turgid mass of commercial and administrative documents, business archives, and inventories. "A huge library discovered ... in Central Anatolia is entirely made up of business transactions; and apart from a solitary text, and that a curse, there is not one of a literary kind" (Sandars 1960, 12). Business, administration, and a curse: here is an epitaph for Western Civilization.

Some form of capitalism, then, has been with us for several thousand years, since cities, agriculture, priesthoods, armies, war, and "business writing" emerged together in the ancient Middle East. Business and state power, represented at the dawn of history by priest-kings recording tribute, have gone hand in hand to the present day, even unto the "state capitalism" of the late USSR and the "Red Capitalism" of contemporary China. With power comes the possibility of oppression. It is noteworthy that the framers of the American constitution, realizing that liberty could never be taken for granted, "assumed a ceaseless and bitter struggle between the interests of the few and the hopes of the many, between those who would limit and those who would extend the authority of the people" (Lapham 1989, 8).

By the mid-twentieth century, Huxley's *Brave New World* (1932) had defined the absurd end point of Western progress, and Orwell's *Animal Farm* (1945) had shown us just how far we should trust our leaders, while his *1984* (1949) was probably the first public perception of the huge power of information technology to rewrite history, control thought, and invade privacy. And there is no shortage of late-twentieth-century critiques of this cruel foolishness – from the critical theory of Herbert Marcuse's (1964) "one-dimensional man" living an administered life of overconsumption, to Ezra Mishan's (1986, 179) convincing economic argument that "more is worse" in a society in which power elites increase popular consumption by constantly feeding the springs of consumer discontent. What Aristotle saw as a problem 2,500 years ago is now normal: an incessant expansion of consumerist "needs" that leaves people perpetually unsatisfied and too preoccupied with trivial choices to take part either in public life or in the real art of living well.

Like some perceptive domicide victims, Seabrook does not hesitate to lay blame: "For what exactly is being held up to our wonder and admiration? The rich are nothing more than the most monstrous predators upon the earth's resources, the cannibalistic devourers of the substance of the poor. Are we seriously expected to applaud their prodigious appetites, their bottomless capacity for using up and spending? They are the most baleful force ever unleashed upon the world.

The ransackers of its beauties, the plunderers of its natural treasures, the greedy and parasitical exhausters of its fragile covering ... There is no problem of poverty ... but for the more intractable problem of wealth" (Seabrook 1988, 21). And in *The Insanity of Normality*, Gruen (1992, 86, 91, 138) accepts Tuchman's view that history is a record of the actions of "realists" who keep leading us to destruction. He notes that our self-esteem depends on success, which is, in turn, defined in terms of control and domination, and he asserts that we despise victims because we hate the victim in ourselves.

Williams (1975, 362) puts the case most succinctly: "The active powers of minority capital, in all its possible forms, are our most active enemies, and ... they will have to be not just persuaded but defeated and superseded." The possibility of defeating such power begins with questioning the insanity of normality. Why do we so readily accept the capitalists' myths that theirs is the most "natural" economic system, the only system that can cope with a complex postmodern world, the system that will best deliver the greatest happiness to the greatest number? Or that capitalism is a system made up of innumerable economic decisions made by numberless decision makers and benefiting hordes of thoughtful stakeholders? The truth is that the major transformative decisions, on which all else depends, are made worldwide by only a few hundred thousand very powerful people in charge of giant national and transnational corporations, who have strong links with politicians, bureaucrats, and the military. An even more unpleasant truth is that today, 500 men control more individual wealth than the combined worth of the poorest two billion people on the planet. Can they all have reached this position without committing systematic injustices, including domicide? In crude cases – such as Saudi Arabia, the Persian Gulf States, Indonesia, and many African countries – it is quite obvious that the operations of the state are dominated by, and work largely for the benefit of, a single family and its associates. Power is more diffused in the Western world, but in Canada, for example, the Business Council on National Issues, a group of chief executive officers of the 150 largest companies in the country, routinely influences political decisions (Finn 1998, 4). Our consent having been manufactured, we regard such injustices as "normal."

In the twenty-first century, human rights will be lost on a global scale. Injustice is normal in the less-developed world (Bales 1999) but even in the developed nations "one of the striking qualities of the post-Cold War globalization is how easily business and government in the capitalist democracies have abandoned the values they putatively espoused for forty years during the struggle against communism. ... Concern for human rights ... has been pushed aside by commercial opportunity" Greider 1997, 37). With globalization, the free market, and the search for international competitiveness and economic efficiency, benefits will accrue to the few, and the hopes of the many will be dashed. Competitiveness, it must be realized, is a zero-sum game in which there are winners and losers. We may look forward in trepidation to a late twenty-first century world when national states will no longer be able to mediate the power of superstate corporations, thus bringing the global corporation into direct contact with the localities in which almost all of us live. Geopolitically, nation states will remain in name only, since the development of supranational organizations and free-trade associations will end the ability of states to protect their citizens. The confrontation of powerful global corporations with vulnerable localities does not look promising for place and home.

The defeat of this growing system, and of the few who benefit from it, will take a concerted effort on the part of both national states and the international alliance of grassroots resistance networks now being set up. The rise of a shadow world government in the shape of the exponential growth of NGOs heralds the creation of an interconnected, popular "embryonic nervous system for our global society, affording a degree of sensitivity unmatched by the nation states and their 2000 intergovernmental agencies, which are too often tongue-tied by diplomacy" (Myers 1984, 238). The goals of such NGOs are to oppose power elites, to act as a global conscience, to invent and implement alternative development strategies, and to broaden the base of power to allow bottom-up development to challenge top-down technocratic approaches. The vision of globalization from above, led by commerce, is thus being challenged by international social activism, *Globalization from Below* (Brecher, Costello, and Smith, 2000).

And further, change will occur inevitably when enough people become conscientized, and millions of individual decisions are made to eschew greed, learn moderation, appreciate simplicity, and, understanding that more is less, live lives *Simple in Means, Rich in Ends* (Devall 1990). This, in turn, requires a restructuring of education that turns away from making a product with exchange value in the market and seeks to empower people in political, artistic, intellectual, and spiritual ways. It is a telling indictment of our civilization that currently "only a very tiny proportion of formal education ... is devoted to the skills involved in the enjoyment of living" (Atchley 1980, 32).

Although the above is a consummation devoutly to be wished, powerful forces are arrayed against any such future. The prospects for an acceleration of domicide in the twenty-first century are excellent insofar as the globe's 200 countries have only two basic forms of government: dictatorship and so-called liberal democracy. In dictatorships –

whether ideological, fundamentalist, or absolutist – domicide can always be justified. The governments of "free-market" democracies, whatever their rhetoric, invariably act to strengthen the position of the rich and powerful, degrade the poor, and weaken the middle class. Such is the interpenetration of politics, business, bureaucracy, and the military in many industrial nations, that we need a more accurate image than former US president Eisenhower's "military-industrial complex." We find this in the notion of the corporate state. The corporate state is an interlocking amalgam of powerful corporate (major transnational companies and conglomerates) and government elites (politicians, bureaucrats, the military). It has surfaced before in the twentieth century in Fascist Italy, Germany, and Japan. At the beginning of the twenty-first century, it is already in place in the world's largest nation, China, and in a somewhat different form, in the world's most powerful nation, the United States.

The goal of the American corporatists, with their Japanese and European allies, is to dominate the global economy and world geopolitics, the motives being power, wealth, and the common good of their shareholders. The means by which this hegemony will be accomplished is called "globalization," which involves tariff-free trade on a global scale, policed by transnational corporations, military interventions, and organizations such as the World Trade Organization, the World Bank, and the International Monetary Fund. The *Multilateral Agreement on Investment*, stalled by global grassroots resistance in 1998, would effectively destroy national sovereignty in favour of transnational corporations and wealthy investors. It is already in place regionally in the form of the *North American Free Trade Agreement*. The outcomes of globalization are already being felt worldwide in massive job losses, the transfer of work offshore, currency collapses, the curtailment of social programs, the growing gap between rich and poor, the destruction of national and regional cultures, the invasion of privacy, and the development of surveillance techniques on earth and from space. When globalization is in full swing, we can look forward to a world controlled by a very few, to the detriment of over six billion people, and in which domicide will be quite normal whenever required by the "magic of the market." The ultimate end point of this logic will be terracide.

The alternative to a chessboard world raped of its resources, homes, and homelands by corporate warriors is to envision, then create, the exact opposite. It is not difficult to imagine a world in which co-operation replaces competition, community provides more satisfaction than consumerism, the local is given social (but not necessarily environmental) priority over the global, the poor as well as the rich are enabled to live lives of self-fulfillment, home and hearth are valued as greatly as cosmopolitanism, and a just sufficiency is preferred to capitalism's orgy of consumption (Brandt 1995, Moore 1996, Cox 1997, *New Internationalist* April 1997). The "humanistic capitalism" envisaged by Harman (1984, 14), in which "humane and social and ecological values predominate over short-term economic considerations," could become more than a vision.

Thousands of community forums, co-operative ventures, local development boards, grassroots organizations of every kind, community development associations, spiritual reawakenings, voluntary simplicity movements, and alternative lifestyles already in place across the planet argue that a bottom-up transformation of the globe is already on the horizon. Just as transnational economic power is currently replacing the nation-state as chief decision maker, so the urge to localness, the belief that small is beautiful, will inexorably bring about, via millions of decisions made by individuals and small groups, a global network of individuated, self-governing localities, micro-homelands in which the centred identity of home is inviolate by uncaring outside forces. The alternative, to put it mildly, is extremely bleak. In the words of Italo Calvino (1972): "The inferno of living is not something that will be; if there is one, it is what is already here, the inferno where we live every day, that we form by being together. There are two ways to escape suffering it. The first is easy for many: accept the inferno and become such a part of it that you can no longer see it. The second is risky and demands constant vigilance and apprehension: seek and learn to recognize who and what, in the midst of the inferno, are not inferno, then make them endure, give them space."

Domicide will be eliminated only if such alternative spaces have the courage to become an ethical platform from which to speak truth to power.

Bibliography

ARCHIVAL MATERIAL

British Columbia, Ministry of Government Services
Archives and Records Service:

Arrow Lakes News (1961) "Editorial" *Arrow Lakes News*. September.

Clough, W. (1961) Letter to the Comptroller of Water Rights for British Columbia. *Record of Proceedings. Hearings Concerning Applications to Build the Mica, Duncan and High Arrow Dams*, Revelstoke, B.C., 26 September.

Comptroller of Water Rights for British Columbia (1961a). Record of Proceedings. Hearings Concerning Applications to Build the Mica, Duncan and High Arrow Dams, Nakusp, B.C. Volume V, September 29, 1961, pp. 339–579.

Comptroller of Water Rights for British Columbia (1961b). Record of Proceedings. Hearings Concerning Applications to Build the Mica, Duncan and High Arrow Dams. Nakusp, B.C.: Volume VI, September 30, 1961, pp. 590–768.

Comptroller of Water Rights for British Columbia (1961c). Record of Proceedings. Hearings Concerning Applications to Build the Mica, Duncan and High Arrow Dams. Castlegar, B.C. Volume VII, October 3, 1961, pp. 769–919.

International Joint Commission (IJC). United States and Canada. (1951) *Record of Public Hearings. Libby Dam and Pen d'Oreille Power Projects, Nelson and Cranbrook*, March. Mimeographed.

Nimsick, L. (1968) Correspondence to Leo Nimsick, M.L.A. held in the British Columbia Archives, MSS 854, Box 4, Files 2 and 4.

OTHER SOURCES

Abbey, E. (1979) *Abbey's Road*. New York: Dutton.

– (1984) "Forward!" 3–5 in D. Foreman and B. Hayward, (eds), *Ecodefence: A Field Guide to Monkey-wrenching*. Chico, Cal.: Abbzuq Press.

– (1988) *One Life at a Time, Please*. Markham, Ontario: Fitzhenry and Whiteside.

Aberley, D. (1993) *The Boundaries of Home: Mapping for Local Empowerment*. Philadelphia: New Society.

Abrams, C. (1971) *The Language of Cities*. New York: Avon.

Ackerly, J. (1993) "A Long, Cold Saga into Exile," *Victoria Times-Colonist* 20 July.

Adams, D. (1979) *The Hitchhiker's Guide to the Galaxy*. New York: Pocket Books.

Adams, P. (1991) *Odious Debts*. London: Earthscan.

– (1993) "Planning for Disaster: China's Three Gorges Dam," *Multinational Monitor*. September:16–20.

Ahrentzen, S. et al. (1989) "Space, Time and Activity in the Home: A Gender Analysis." *Journal of Environmental Psychology* 9: 89–101.

Aiken, L. (1985) *Dying, Death and Bereavement*. Boston: Allyn and Bacon.

Aitkenhead, D. (1996) "First it was Tunnels, now it's Runways," *Independent on Sunday* 2 February.

Alain-Fournier (1966) *Le Grand Meaulnes*. New York: Penguin.

Alladin, I. (1993) *Economic Miracle in the Indian Ocean: Can Mauritius Show the Way?* Stanley, Rose Hill, Mauritius: Editions de l'Ocean Indien.

Allison, K. (1970) *Deserted Villages*. Toronto: McMillan.

Altman, I. (1992) "Place Attachment and Interpersonal Relationships," in *Socio-Environmental Metamorphoses*. Proceedings of IAPS Conference, Thessaloniki, Greece.

Altman, I., and M. Chemers (1980) *Culture and Environment*. Monterey, California: Brooks/Cole.

Altman, I., and M. Gauvain (1981) "A Cross-cultural and Dialectic Analysis of Homes," 283–319 in L. Liben et al. (eds.), *Spatial Representations and Behaviour Across the Life-Span*. New York: Academic Press.

Amarteifo, G. (1970) "Social Welfare," 103–47 in R. Chambers (ed.), *The Volta Resettlement Experience*. New York: Praeger Publishers.

Ambrose, P., and Colenutt, R. (1975) *The Property Machine*. London: Penguin.

Amte, B. (1990) *The Case Against Narmada and the Alternative Perspective*. Anandwan, Maharashtra, India: Baba Amte Movement against Big Dams.

Anderson, M. (1964) *The Federal Bulldozer*. Cambridge, Massachusetts: M.I.T. Press.

Angelo, M. (1996) "Nepal Reverses Mega-Dam Trend," *Victoria Times-Colonist* 2 February.

Anonymous (1972) "Nuevo, Home, Capitalism and Geography," *Place* 2: 131–32.

Anonymous (1997) *Home Is Where the Heart Is*. Oxford: Past Times.

Anson, B. (1981) *I'll Fight You for It*. London: Jonathan Cape.

Appleyard, D. (1979) "Home," *Architectural Association Quarterly* 2: 4–20.

Arnstein, S. (1969) "A Ladder of Citizen Participation," *AIP Journal* 35: 216–24.

Ashini, D. (1996) Foreword to M. Wadden, *Nitassinan*. Vancouver: Douglas and McIntyre.

Atchley, R. (1980) *The Social Forces in Later Life*. Belmont, California: Wadsworth.

Australian Agency for International Development. (1997) *Economic Survey of Papua New Guinea*. Canberra: Commonwealth of Australia.

Bailey, C. (1994) "Eviction in the Desert," *Toronto Globe and Mail* 3 January.

Bailey, I. (1993) "Newfoundland's Junked Resettlement Idea Reviving," *Victoria Times-Colonist* 23 September.

Bair, D. (1978) *Samuel Beckett*. New York: Harcourt, Brace.

Baird, V. (ed.) (1999) *The Little Book of Big Ideas*. Oxford: New Internationalist.

Baker, A., and Biger, G. (1992) *Ideology and Landscape in Historical Perspective*. Cambridge: Cambridge University Press.

Bales, K. (1999) *Disposable People: New Slavery in the Global Economy*. Berkeley, CA: University of California Press.

Balian, C. (1985) "The Carson Productions Interview" 58–61 in J. Hepworth and G. McNamee (eds) *Resist Much, Obey Little: Some Notes on Edward Abbey*. Salt Lake City: Dream Garden Press.

Bancoult, O. (2000) "Lies told over Diego Garcia," *Guardian Weekly* 22 November.

Barber, M., and G. Ryder (eds.) (1990) *Damming the Three Gorges: What Dam Builders Don't Want You to Know*. (2nd ed.) London: Earthscan Publications.

Bardwick, J. (1991) *Danger in the Comfort Zone*. New York: American Management Association.

Barfoot, R. (1971) "On Second Thoughts," *The Architect* 1:33.

Barnabas, A., and M. Bartolemé (1973) *Hydraulic Development and Ethnocide: The Mazatec and Chinantec People of Oaxaca, Mexico*. Copenhagen: International Working Group for Indigenous Affairs.

Barrett, A. (1992) *The Forms of Water*. New York: Washington Square Press.

Baum, A. (1987) *Cataclysms, Crises and Catastrophes: Psychology in Action*. Washington: Americal Psychological Association.

Bell-Fialkoff, A. (1993) "A Brief History of Ethnic Cleansing," *Foreign Affairs* 72: 110–21.

Belloc, H. (1912) *The Servile State*. London: T.N. Foulis.

Benedictus, D. (1983) *Local Hero*. London: Penguin.

Benjamin, D. (1995) *The Home*. Aldershot, U.K.: Avebury.

Beresford, M. (1955) "The Lost Villages of Yorkshire," *The Yorkshire Archaeological Journal* 38: 44–70.

Berg, L. (1990) *Aboriginal People, Aboriginal Rights, and Protected Areas: An Investigation of the Relationship between the Nuu-chah-nulth People and the Pacific Rim National Park Reserve*. Victoria, British Columbia: University of Victoria unpublished Master of Arts thesis.

Berger, J. (1984) *And Our Faces, My Heart, Brief as Photos*. New York: Pantheon.

– (1990) "You Can't Go Home Again: The Hidden Pain of Twentieth Century Life," *Utne Reader* 39: 85–87.

Berger, T. (1977) *Northern Frontier, Northern Homeland: The Report of the Mackenzie Valley Pipeline Inquiry*. Ottawa: Supply and Services Canada.

– (1986) "The Probable Economic Impact of a Mackenzie Valley Pipeline," 179–93 in J. Ponting, (ed.), *Arduous Journey*. Toronto: McClelland and Stewart.

Berman, M. (1982) *All that is Solid Melts into Air*. New York: Simon and Schuster.

Bishop, E. (1976) *Geography III*. New York: Farrar, Straus and Giroux.

Black, R., and V. Robinson (1993) *Geography and Refugees*. London: Belhaven.

Blaut, J. (1993) *The Colonizer's Model of the World*. New York: Guilford.

Blowers, A. (1980) *The Limits of Power*. Oxford: Pergamon Press.

Blythe, R. (1969) *Akenfield*. Harmondsworth, Middlesex: Penguin.

– (1980) *The View in Winter: Reflections on Old Age*. Harmondsworth, Middlesex: Penguin.

– (1981) "The Greening of Suburbia," *The Guardian* 6 December, 22.

Bluestone, B., and B. Harrison (1980) *Capital and Communities: The Causes and Consequences of Private Disinvestment*. Washington: The Progressive Alliance.

Bollnow, O. (1960) "Lived-Space," *Philosophy Today* 5: 31–9.

Bordessa, R. (1989) "Between House and Home: The Ambiguity of Sojourn," *Terra* 101: 34–7.

Borneman, K. (1968) Notes taken at public meeting re Lake Kookanusa Reservoir preparation (19 September).

– (1993) Discussion regarding files and notes from the period of clearing for the Lake Kookanusa Reservoir, 1968 (personal discussion).

Boschetti, M. (1990) "Reflections on Home: Implications for Housing Design for Elderly Persons," *Housing and Society* 17: 57–65.

– (1995) "Attachment to Personal Possessions: An Interpretive Study of the Older Person's Experience," *Journal of Interior Design* 21:1–12.

Bothwell, S., R. Gindroz, and R. Lang (1998) "Restoring Community through Traditional Neighbourhood Design: A Case Study of Diggs Town Public Housing," *Housing Policy Debate* 9: 89–114.

Bourdieu, P., and A. Sayad (1964) *Le Déracinement: La Crise d'Agriculture traditionnel en Algerie*. Paris: Editions Nesle.

Bowen, E. (1986) [originally 1938] *The Death of Heart*. London: Penguin

Bowles, R. (1981) *Social Impact Assessment in Small Communities*. Toronto: Butterworths.

– (1982) *Little Communities and Big Industries*. Toronto: Butterworths.

Bradbury, J., and I. St. Martin (1983) "Winding Down in a Quebec Mining Town: A Case-Study of Schefferville," *Canadian Geographer* 27: 128–44.

Bradbury, J., and M. Sendbuehler (1988) "Restructuring Asbestos Mining in Western Canada," *The Canadian Geographer* 32: 296–306.

Bradsaw, R. (2000) "In Pursuit of the Magic Ring," *Globe and Mail*, 10 January.

Brandt, B. (1995) *Whole Life Economics*. Philadelphia: New Society.

Bray, M. and A. Thompson (eds.) (1992) *At the End of the Shift*. Toronto: Dundurn Press.

Brazier, C. (1996) "Join the Resistance" and "In the Dock" *New Internationalist* January: 7–11; May: 7–8.

Brecher, J., T. Costello and B. Smith (2000) *Globalization from Below: The Power of Solidarity*. Cambridge, Mass.: South End Press.

Breitbart, M., and R. Peet (1973) "A Critique of Advocacy Planning," 97–107 in D. Ley (1974) *Community Participation and the Spatial Order of the City*. Vancouver: Tantalus Research.

Brink, A. (1974) *Looking on Darkness*. London: Allen.

British Columbia. *Water Act, R.S.B.C 1979, c.429*. Victoria, British Columbia: Queen's Printer.

– Department of Lands, Forests and Water Resources. (1962) Order. Water Act Section 15, (File 0236915, December 14, 1962) and Conditional Water Licence No. 27066, (File No. 0236915, April 16, 1963).

– Downstream Benefits Steering Committee. (1993a) "The Canadian Entitlement," *Columbia Report*. Victoria, British Columbia: Ministry of Energy, Mines and Petroleum Resources.

– Downstream Benefits Steering Committee. (1993b) "The Columbia River Treaty," *Columbia Report*. Victoria, British Columbia: Ministry of Energy, Mines and Petroleum Resources.

– Downstream Benefits Steering Committee. (1993c) "B.C. Reaches Agreement with the U.S. on Entitlement Delivery," *Columbia Report*. Victoria, British Columbia: Ministry of Energy, Mines and Petroleum Resources.

British Columbia Hydro and Power Authority. (1961) *Progress*. Vancouver, British Columbia: BC Hydro.

– (1964) *Property Owners' Guide*. Vancouver, British Columbia: BC Hydro.

– (1965) *The New Outlook for the Arrow Lakes*. Vancouver, British Columbia: BC Hydro.

– (1966) Columbia River Treaty Projects. Assistance to Individuals, Organizations and Communities in the Reservoir Area. Mimeographed.

– (n.d.) *Columbia Construction Progress. Arrow Project. Review of Construction*. Vancouver, British Columbia: BC Hydro.

British Columbia Water Resources Service. (Draft, 1968 and Brochure, n.d.) *Property Owners' Guide. Libby Dam Reservoir Area*. Province of British Columbia: Department of Lands, Forests and Water Resources.

– (1968) *Progress Report*. [on Lake Kookanusa Reservoir preparation] 19 September.

– (1969) *Libby Reservoir Preparation*. File No. 0178369A, March 10.

– (1972–1981) *Libby Reservoir Preparation. Highways Land Acquisition*. File No. 0178369-"R". Vol. 1 on Microfilm, Volume 2 on File.

– (n.d.) *The Libby Dam Reservoir*. Province of British Columbia: Department of Lands, Forests and Water Resources.

Brody, H. (1988) *Maps and Dreams: Indians and the British Columbia Frontier*. Vancouver: Douglas and McIntyre.

Bromley, D. (1991) *Environment and Economy: Property Rights and Public Policy*. Oxford: Blackwell.

Bronskill, J. (1993) "Relocating Community Considered as Remedy," *Victoria Times-Colonist* 3 February.

Brontë, C. (1971) *Jane Eyre*. New York: W.W. Norton.

Brooker, E. (1993) "To Stop the Road, First Remove Dentures," *Weekend Telegraph* 15 May.

Brookner, A. (1987) *A Friend from England*. New York: Pantheon.

Brookshire, D., and R. D'Arge (1980) "Adjustment Issues of Impacted Communities or, Are Boomtowns Bad?," *Natural Resources Journal* 20: 523–46.

Brown, G. (1972) *Greenvoe: A Novel*. London: Penguin.

Brown, L. (1997) *State of the World 1997*. New York: Norton.

Bryson, B. (1988) "Fat Girls in Des Moines," *Granta* 23: 23–43.

Bugslag, B. (1993) Discussion regarding clearing for the Lake Kookanusa reservoir, 1968.

Burgel, G. (1992) "The Big City: Audacity and Necessity," 1–8 in *Socio-Environmental Metamorphoses*. Proceedings of the IAPS Conference, Thessaloniki, Greece.

Burton, I. (1994) "Deconstructing Adaptation, and Reconstructing," *Delta: Newsletter of the Canadian Global Change Program*. 5: 14–15.

Burton, I., et al. (1978) *The Environment as Hazard*. New York: Guilford.

Buttimer, A. (1976) "Grasping the Dynamism of Lifeworld," *Annals, AAG* 66: 2–8.

– (1980) "Home, Reach and the Sense of Place," 166–87 in A. Buttimer and D. Seamon (eds.), *The Human Experience of Space and Place*. London: Croom Helm.

Bygrave, M. (1974) *About Time*. London: Quartet.

Calabresi, G. and P. Bobbitt (1978) *Tragic Choices*. New York: W.W. Norton.

Calhoun, J. (1997) "Rosewood," *Theatre Crafts International*. 31: 40–3.

Calvino, I. (1972) *Invisible Cities*. London: Penguin.

Canada. House of Commons. (1964) *Debates, Official Record*. Ottawa: Queen's Printer, 5 March, p. 599ff.

– Statutes. (1955) 3–4 Eliz. II, ch. 47, *The International River Improvements Act*. Ottawa: Queens Printer.

Canadian Broadcasting Corporation (1994) "Gravelgate," Primetime News Magazine. 3 May.

Canadian Centre for Policy Alternatives *Monitor* September 1998.

Cant, G., et al. (1993) *Indigenous Land Rights in Commonwealth Countries*. Christchurch, New Zealand: Department of Geography, University of Canterbury, New Zealand.

Caro, R. (1974) *The Power Broker: Robert Moses and the Fall of New York*. New York: Knopf.

Carr, E. (1942) *The Book of Small*. Toronto: Clarke Irwin.

Carter, A. (1982) *Nothing Sacred*. London: Virago Press.

Carter, S. (1976) *Cherokee Sunset*. New York: Doubleday.

Castells, M. (1989) *The Informational City*. Oxford: Blackwell.

Catanese, A. (1984) *The Politics of Planning and Development*. Beverly Hills, California: Sage.

Chambers, J. and G. Mingay (1966) *The Agricultural Revolution 1750–1880*. London: Batsford.

Chambers, R. (1969) *Settlement Schemes in Tropical Africa*. London: Unwin.

– (ed.) (1970) *The Volta Resettlement Experience*. New York: Praeger Publishers.

Charette, C. (1987) "Homelessness: Community Perspectives," *Institute of Urban Studies Newsletter* 22: 4.

Charlie, M. (1993) Presentation in Victoria, B.C., on 26 October at First United Church.

Chawla, L. (1994) *In the First Country of Places: Nature, Poetry and Childhood Memory*. Albany, New York: SUNY Press.

– (1995) "Reaching Home: Reflections on Environmental Autobiography," *Environmental and Architectural Phenomenology Newsletter* 6: 12–15.

Chicago Tribune (1989) "Foreign Doublespeak," *American Journal of Doublespeak* 16: 2.

Christopher, A. (1991) "Changing Patterns of Group-area Proclamations in South Africa 1950–1989," *Political Geography* 10: 240–53.

– (1994) "South Africa: The Case of a Failed State Partition," *Political Geography* 13: 123–36.

Clairmont, D., and D. Magill (1987) *Africville*. Toronto: Canadian Scholars' Press.

Clark, M., and J. Herington. (1988) *The Role of Environmental Impact Assessment in the Planning Process*. London: Mansell.

Clodumar, K. (1994) "Plenary Address," Global Conference on the Sustainable Development of Small Island Developing States, Barbados 25 April–6 May.

Coelho, G., and P. Ahmed (1980) *Uprooting and Development: Dilemmas of Coping with Modernization*. New York: Plenum Press.

Cohen, B. (1990) *The Vietnam Guidebook*. New York: Harper and Row.

Cohen, R. (1994) *International Protection for Internally Displaced Persons – Next Steps*. Washington, D.C.: Refugee Policy Group.

Cohen, R., and F. Ahearn, Jr. (1980) *Handbook for Mental Health Care of Disaster Victims*. Baltimore: The Johns Hopkins University Press.

Cohen, R., and F. Deng (1998) "Exodus within Borders," *Foreign Affairs* 77:12–16.

Cole, J. (1987) *Crossroads, the Politics of Reform and Repression, 1976–1986*. Johannesburg, South Africa: Ravan Press.

Colson, E. (1971) *The Social Consequences of Resettlement: The Impact of the Kariba Resettlement Upon the Gwembe Tonga*. Manchester: University of Manchester Press.

Columbia Basin Trust (1998a) "Trust Funds Dispute Resolution Process to Help Resolve Claims due to Dam Construction," *Columbia Basin Trust News Release*. Nakusp, B.C.: Columbia Basin Trust.

– (1998b) *Columbia Basin Trust Annual Report*. Nakusp, B.C.: Columbia Basin Trust.

Columbia River Treaty Committee (1993) *Community Involvement in the Columbia*. Victoria, British Columbia: Ministry of Energy, Mines and Petroleum Resources.

Commission on the Third London Airport (1971) *Report*. London: Her Majesty's Stationery Office.

Comptroller and Auditor General, U.K. (1994) *Department of Transport: Acquisition, Management and Disposal of Lands and Property Purchased for Road Construction*. London: Her Majesty's Stationery Office.

Comstock, D., and R. Fox (1982) *Participatory Research as Critical Theory: The North Bonneville, USA Experience*. Pullman, Washington: Washington State University Department of Sociology, Photocopy.

Cooper, C. (1974) "The House as Symbol of the Self," 130–46 in J. Lang et al. (eds.) *Design for Human Behaviour: Architecture and the Behavioural Sciences*. Stroudsberg, Pennsylvania: Dowden, Hutchinson and Ross.

Cooper-Marcus, C. (1995) *House as a Mirror of the Self*. Berkeley, California: Conari.

Copes, P. (1972) *The Resettlement of Fishing Communities in Newfoundland*. Ottawa: Canadian Council on Rural Development.

Cornaton, M. (1967) *Les Regroupements de la Décolonization en Algerie*. Paris: Editions Nesle.

Cornerstone Planning Consultants (1994) *1994 Columbia-Kootenay Symposium Saturday Night Report*. Photocopy.

Cox, K. (ed.) (1977) *Spaces of Globalization Reasserting the Power of the Local*. New York: Guilford.

Cragg, B. (1982) "Wild Daniel's Farm: A Family Geography," *Landscape* 26: 41–8.

Cramer, R. (1960) "Images of Home," *Journal of the American Institute of Architects* 46: 40–9.

Creighton-Kelly, C. (1992) "Culture Talks," *Boulevard* Autumn: 18.

Crowfoot, J., and J. Wondolleck (eds.) (1990) *Environmental Disputes: Community Involvement in Conflict Resolution*. Washington, D.C.: Island Press.

Cruickshank, J. (1987) "Razing of Village Leaves Bitter Legacy for Indians," *Toronto Globe and Mail* 22 September, 1.

Csikszentmihalyi, M., and E. Rochberg-Halton (1981) *The Meaning of Things: Domestic Symbols and the Self*. Cambridge: Cambridge University Press.

Cummings, A. (1972) "The Bougainville Copper Industry: Construction Phase," *Australian Geographer* 12: 55–6.

Cummings, B. (1990) *Dam the Rivers, Damn the People*. London: Earthscan Publications.

Cutter, S. (1993) *Living with Risk: The Geography of Technological Hazards*. London: Arnold.

Cybriwsky, R. (1995) Review of *Our Land Was a Forest* by S. Kayano. *Professional Geographer* 47: 230.

Dai Qing. (1992) *Yangtze! Yangtze! Debate over the Three Gorges Project*. London: Earthscan Publications.

Dalrymple, W. (1997) *From the Holy Mountain*. London: Harper Collins.

Daniels, A. (1991) *Utopias Elsewhere: Journeys in a Vanishing World*. New York: Crown.

Davidoff, P. (1965) "Advocacy Planning and Pluralism in Planning," *Journal of the American Institute of Planners* 31: 331–7.

Davies, J. (1972) *The Evangelistic Bureaucrat*. London: Tavistock.

Davison, J. (1980) *The Fall of a Doll's House: Three Generations of American Women and the Houses They Lived In*. New York: Holt, Rinehart and Winston.

Deak, I. (1989) "The Incomprehensible Holocaust," *The New York Review of Books* 28 September, 63–72.

Dear, M., and S. Flusty (1998) "Postmodern Urbanism," *Annals of the Association of American Geographers* 88: 50–72.

Dearden, P., and L. Berg (1993) "Canada's National Parks: A Model of Administrative Penetration," *Canadian Geographer* 37: 194–211.

Deats, R. (1997) "The Global Spread of Active Nonviolence," *Ploughshares Monitor* 18: 11–16.

Denman, D. (1978) *The Place of Property*. Berkamstead, Hertfordshire: Geographical Publications.

Denoon, D., et al. (eds.) (1997) *The Cambridge History of the Pacific Islanders*. Cambridge: Cambridge University Press.

Depres, C. (1991) *Form, Experience and Meaning of Home in Shared Housing*. Milwaukee: Ph.D. dissertation, University of Wisconsin.

Desmond, C. (1971) *The Discarded People: An Account of African Resettlement in South Africa*. Harmondsworth, Middlesex: Penguin.

DeSpelder, L., and A. Strickland (1987) *The Last Dance: Encountering Death and Dying*. Mountain View, California: Mayfield.

Deval, W. (1990) *Simple in Means, Rich in Ends: Practising Deep Ecology*. London: Green Print.

DeVilliers, M., and S. Hirtle (1997) *Into Africa: A Journey Through the Ancient Empires*. Toronto: Key Porter.

Dewey, J. (1946) *The Public and Its Problems: An Essay in Political Inquiry*. Chicago: Gateway.

Dickman, P. (1973) "Spatial Change and Relocation," 145–74 in J. Rogge (ed.) *Developing the Subarctic*. Winnipeg, Manitoba: The University of Manitoba Department of Geography.

Dinesen, I. (1972) *Out of Africa*. New York: Vintage Books.

Dixon, N. (1987) *Our Own Worst Enemy*. London: Cape.

Dodds, B. (1989) "Letter from Chernobyl," *Peace* 5: 6, 7, 24, 25.

Domhoff, G. (1978) *Who Really Governs? New Haven and Community Power Reexamined*. New Brunswick, New Jersey: Transaction Books.

Dorcey, A. (1998) "World Commission on Dams," *Water News* 17, 25–6.

Douglas, M. (1991) "A Kind of Space," 287–307 in A. Mack (ed.), *Home: A Place in the World*. Special Edition of *Social Research*. New York: New School for Social Research.

Dovey, K. (1978) "Home as an Ordering Principle in Space," *Landscape* 22: 27–30.

– (1985) "Home and Homelessness," 33–61 in I. Altman and C. Werner (eds.) *Home Environments*. New York: Plenum Press.

Drum, R. (1998) Letter, *Environmental and Architectural Phenomenology Newsletter* 9, 3.

Dugard, J. (1985) "Denationalization: Apartheid's Ultimate Plan," 17 in Platzky, L., and C. Walker (1985) *The Surplus People, Forced Removals in South Africa*. Johannesburg: Ravan Press.

Dwyer, A. (1992) "The Trouble at Great Whale," *Equinox* 61: 28–41.

Dyer, G. (1985) *War*. New York: Crown.

Ebaugh, H. (1988) *Becoming an EX*. Chicago: University of Chicago Press.

Ebrahim, M. (1984) "Nomadism, Settlement and Development," *Habitat International* 8: 125–41.

Economist April 1997.

Egenter, N. (1992) "O.F. Bollnow and the Ontology of Home and Movement Outside," Paper prepared for the symposium on *The Ancient Home and the Modern Internationalized Home: Dwelling in Scandinavia.* University of Trondheim, Norway, 20–23 August 1992.

Ekberg, S., et al. (1972) "Task Force Report," 104–15 in E. Browning and D. Forman (eds.), *The Wasted Nations.* New York: Harper Colophon Books.

Ekins, P. (1992) *A New World Order. Grassroots Movements for Global Change.* London: Routledge.

Eliade, M. (1957) *The Sacred and the Profane: The Nature of Religion.* New York: Harcourt, Brace and World.

Eliot, J., et al. (1993) *Thailand, Indochina and Burma Handbook.* Bristol, U.K.: Trade and Travel.

Elliott, H. (1993) "Runway Goes Nowhere Fast," *The Times* 29 July.

Ellis, W., and J. Blair. (1986) "Bikini: A Way of Life Lost," *National Geographic* 169: 813–34.

Ellwood, W. (1995) "Nomads at the Crossroads," *New Internationalist* April: 7–10.

Elmendorp, E. (1990) "Return to the Lost Villages" *Guardian Weekly* 17: 21.

Engel, J. (1983) *Sacred Sands.* Middletown, Conn.: Wesleyan University Press.

Erickson, K. (1976) "Loss of Community at Buffalo Creek," *American Journal of Psychiatry* 133: 302–5.

Evans, J. (1993) "Renewing the Map of Old Familiar Places," *Guardian Weekly* 7 March.

Evans, P. (1997) "Runway to Destruction," *Guardian Weekly* 6 July, 24.

Evenden, L., and I. Anderson (1972) "The Presence of a Past Community: Tashme, British Columbia," 41–66 in J. Minghi (ed.) *Peoples of the Living Land.* Vancouver: B.C. Geographical Series 15.

Eyles, J., and D. Smith (eds.) (1988) *Qualitative Methods in Geography.* Cambridge: Polity Press.

Faegre, T. (1979) *Tents.* New York: Anchor Books.

Fagence, M. (1977) *Citizen Participation in Planning.* Oxford: Pergamon Press.

Fahim, H. (1983) *Egyptian Nubians. Resettlement and Years of Coping.* Salt Lake City: University of Utah Press.

Falah, G. (1996) "The 1948 Israeli-Palestinian War and Its Aftermath: The Transformation and De-Signification of Palestine's Cultural Landscape," *Annals of the Association of American Geographers* 86: 256–85.

Fanning, O. (1975) *Citizen Action.* New York: Harper and Row.

Farrow, M. (1971) "Angry Ranchers Blast Gov't," *The Vancouver Sun* 4 March.

– (1972) "Rancher Vows He'll Stay Put until Dam Floods His Land," *The Vancouver Sun* 26 December.

Fearnside, P. (1993) "Resettlement Plans for China's Three Gorges Dam," 34–58 in M. Barber and G. Ryder *Damming the Three Gorges*. London: Earthscan Publications.

Felt, P. (1977) "National Parks as a Development Tool in Atlantic Canada: A Review of Some Basic Questions," 67–77 in N. Ridler (ed.), *Issues in Regional/Urban Development of Atlantic Canada*. University of New Brunswick at St. John: Social Science Monograph Series II.

Fenton, J. (1983) *The Memory of War and Children in Exile. Poems 1968–1983*. Harmondsworth, Middlesex: Penguin.

Fernie Free Press (1968) "'Fair, Honest Treatment' Promised Flood Area Ranchers," *Fernie Free Press*, 12 September.

Finn, E. (1998) "Provinces following the BCNI's Decentralization Script," *Canadian Centre for Policy Alternatives Monitor* 5: 4.

Finsterbusch, K. (1980) *Understanding Social Impacts*. Beverly Hills, Cal.: Sage Publications.

Finsterbusch, K., et al. (1983) *Social Impact Assessment Methods*. Beverly Hills, Cal.: Sage Publications.

Fisher, D. (1990) *Fire and Ice*. New York: Harper & Row.

Fishman, R. (1987) *Bourgeois Utopias: The Rise and Fall of Suburbia*. New York: Basic Books.

Foster, H. (1976) "Assuming Disaster Magnitude: A Social Science Approach," *Professional Geographer* 28: 241–7.

– (1980) *Disaster Planning: The Preservation of Life and Property*. New York: Springer-Verlag.

Francis, M. (1998) "The Peace Doctors," *Focus: Australia's Overseas Aid Program* July: 10–14.

Frankel, G. (1989) "Army Destroys Gaza Strip Houses," *Guardian Weekly* 26 March: 19.

French, P. (1998) "A Culture of Poverty," *The Sunday Times* 16 August.

Fried, M. (1966) "Grieving for a Lost Home: Psychological Costs of Relocation," 359–79 in J. Wilson (ed.), *Urban Renewal: The Record and the Controversy*. Cambridge, Massachusetts: M.I.T. Press.

– (1982) "Residential Attachment: Sources of Residential and Community Satisfaction," *Journal of Social Issues* 38: 107–19.

Fried, M., and P. Gleicher. (1961) "Some Sources of Residential Satisfaction in the Urban Slum," *Journal of the American Institute of Planners* 27: 305–15.

Friedman, R. (1989) "The Settlers," *The New York Review of Books* 15 June: 14, 56.

Friedmann, J. (1987) *Planning in the Public Domain: From Knowledge to Action*. Princeton, N.J.: Princeton University Press.

Fuentes, C. (1987) Introduction to *The City Builder* by G. Konrad. New York: Penguin, vii–xxv.

Gale, S., and E. Moore (eds). (1975) *The Manipulated City*. Chicago: Maaroufa.

Gallaher, W. (1993) *The Power of Place: How Our Surroundings Shape Our Thoughts, Emotions, and Actions*. New York: Poseidon.

Gallaher, Jr., A. and H. Padfield (eds.) (1980) *The Dying Community*. Albuquerque, N.M.: University of New Mexico Press.

Gans, H. (1962) *The Urban Villagers*. New York: Free Press.

– (1965) "The Failure of Urban Renewal," *Commentary* 39:29–37.

– (1967) *The Levittowners: Ways of Life and Politics in a New Suburban Community*. New York: Random House.

– (1972) *People and Plans*. London: Penguin.

Gibbs, L. (1998) *Love Canal: The Story Continues*. Philadelphia: New Society.

Gibran, K. (1965) *The Prophet*. New York: Knopf.

Gilbert, A. (1989) "The New Regional Geography in English and French-Speaking Countries," *Progress in Human Geography*. 4: 208–28.

Gilbert, M. (1996) *Jerusalem in the Twentieth Century*. New York: Wiley.

Gimson, M. (1980) "Everybody's Doing It," 206–19 in N. Wates and C. Wolmar (eds.) *Squatting, the Real Story*. London: Blackrose Press.

Glavin, T. (1987) "Officials Discover Remote Indian Band," *Vancouver Sun* 12 June.

Gleeson, B. (1994) "The Commodification of Resource Consent in New Zealand," *The New Zealand Geographer* 51: 42–8.

Gleser, G. et al. (1981) *Prolonged Psychosocial Effects of a Disaster*. New York: Academic Press.

Godden, J. and R. (1966) *Two Under the Indian Sun*. New York: Beach Tree Books.

Godkin, M. (1980) "Identity and Place: Clinical Applications Based on Notions of Rootedness and Uprootedness," 73–85 in A. Buttimer and D. Seamon (eds.), *The Human Experience of Space and Place*. London: Croom Helm.

Gold, J., and J. Burgess (eds.) (1982) *Valued Environments*. London: Allen and Unwin.

Goldsmith, E., and N. Hildyard. (1984) *The Social and Environmental Effects of Large Dams*. San Francisco: Sierra Club Books.

Goldsmith, O. (1996) *The Deserted Village*. London: Phoenix.

Goodman, R. (1972) *After the Planners*. London: Penguin.

Goodwin-Gill, G. (1993) "UNHCR and International Protection: Old Problems, New Directions," 14–19 in *World Refugee Survey 1993*. Washington, D.C.: U.S. Committee for Refugees.

Gosnell Sr., J. (1994) "Land Claims: A Two-Way Street," *Victoria Times-Colonist* 22 October, A5.

Goulding, J. (1982) *The Last Outport: Newfoundland in Crisis*. Toronto: Sisyphus Press.

Graber, L. (1976) *Wilderness as Sacred Space*. Washington, D.C.: Association of American Geographers.

Grahame, K. (1908) *Wind in the Willows*. New York: Viking Press.

Gramsci, A. (1971) *Selections from the Prison Notebooks*. New York: International Publishers.

Gray, J. (1997) "Melting Pot Cold in Whitedog," *Globe and Mail* 8 September, A6.

Greenbie, B. (1981) *Spaces: Dimensions of the Human Landscape*. New Haven: Yale University Press.

Greer, G. (1993) "The Adopted Home Is Never Home," *The Guardian* 9 September.

Gregson, N., and M. Lowe (1995) "Home-Making: On the Spatiality of Daily Social Reproduction in Contemporary Middle-Class Britain," *Transactions, Institute of British Geographers* NS 20: 224–35.

Greider, W. (1997) *One World, Ready or Not: The Manic Logic of Global Capitalism*. New York: Simon and Schuster.

Griffiths, L., et al. (1988) *Privy to Privatization: Housing under the Hammer*. Castleford, West Yorkshire: Yorkshire Art Circus.

Grossman, D. (1993) *Sleeping on a Wire: Conversations with Palestinians in Israel*. London: Cape.

Gruen, A. (1992) *The Insanity of Normality*. New York: Grove Weidenfeld.

Gunn, J. (n.d.) *Home?* Victoria: University of Victoria Department of Geography unpublished paper.

Gurr, A. (1981) *Writers in Exile, The Identity of Home in Modern Literature*. Sussex: The Harvester Press.

Guterson, D. (1992) "No Place Like Home," *Harper's Magazine* 285: 55–64.

Gwyn, R. (1992) "Family Movement Grows in U.S.," *Victoria Times-Colonist* 28 December.

Haines, S. (2000) *The Systems Thinking Approach to Strategic Planning and Management*. Boca Raton, Fl.: St. Lucie Press.

Hall, P. (1980) *Great Planning Disasters*. London: Weidenfeld and Nicolson.

Hall, S. (1987) *The Fourth World: The Heritage of the Arctic and Its Destruction*. New York: Vintage.

Halloran, M. (1974) Film "The Reckoning" produced for CBC Television.

Hambleton, A. (1994) Letter, 20 September.

Hanbury-Tenison, R. (1991) *Worlds Apart: An Explorer's Life*. London: Arrow.

Hancox, J. (1992) "Still Clearing the Scottish Highlands," *Guardian Weekly* 26 April, 23.

Hanlon, J. (1984) *Mozambique: The Revolution under Fire*. London: Zed Books.

Harding, J. (1993) *The Fate of Africa: Trial by Fire*. New York: Simon and Schuster.

Hardyment, C. (1990) "A Roam from Home," *Weekend Guardian* 19–20 May, 12.

Harlow, J. (1993) "Thousands Threatened by New Air Terminal," *The Sunday Times* 11 July, 9.

Harman, W. (1984) "Key Choices," 7–20 in D. Korten and R. Klauss (eds.), *People Centered Development*. West Hartford, Conn.: Kermanian Press.

Harper, P., and L. Fullerton (1994) *Philippines Handbook* Chico, Cal.: Moon.

Harrison, R. (1997) "Bare-Faced Cheek," *New Internationalist* April: 26–27.

Har-Shefi, Y. (1980) *Beyond the Gunsights*. Boston: Houghton Mifflin.

Hart, R. (1979) *Children's Experience of Place*. New York: Irvington.

Hartman, C. (1966) "The Housing of Relocated Families," 293–355 in J. Wilson (ed.) *Urban Renewal: The Record and the Controversy*. Cambridge, Massachusetts: M.I.T. Press.

Hartmann, F. (1992) "The Sinister Ideology of 'Ethnic Cleansing'," *The Guardian Weekly* 13 September, 18, 21.

Hartnett, K. (1970) *Encounter on Urban Environment*. Ottawa: Canadian Broadcasting Corporation.

Hartshorne, R. (1949) *The Nature of Geography. A Critical Survey of Current Thought in the Light of the Past*. Lancaster, Penn.: Annals of the Association of American Geographers.

Harvey, D. (1989) *The Condition of Postmodernity*. Oxford: Blackwell.

Hauxwell, H. and B. Cockcroft (1989) *Seasons of My Life*. London: Century Hutchinson.

Havel, V. (1991) "On Home," *New York Review of Books* 5 December, 49.

Hay, R. (1987) "Senses of Place: Experiences from the Cowichan Valley," in R. Le Heron et al., *Geography and Society in a Global Context*. Proceedings 14th New Zealand Geography Conference.

Hayward, D. (1975) "Home: an Environmental and Psychological Concept," *Landscape* 20: 2–9.

Hayward, D., et al. (1976) "The Meanings of Home in Relation to Environmental and Psychological Issues," in A. Weidmann and B. Anderson (eds.), *Priorities for Environmental Design Research*. EDRA 8: 418–20.

Heidegger, M. (1964) *Basic Writings*. New York: Harper & Row.

Hewitt, K. (1983a) *Interpretations of Calamity*. Boston: Allen and Unwin.

– (1983b) "Place Annihilation: Area Bombing and the Fate of Urban Places," *Annals of the American Association of Geographers* 73: 257–84.

– (1987) "The Social Space of Terror: Towards a Civilian Interpretation of Total War," *Society and Space, Environment and Planning D* 5: 445–74.

– (1993) "Reign of Fire," 25–46 in J. Nipper and M. Nutz (eds.) *Kriegszerstorung und Wiederaufbau deutscher Städte*. Cologne: Kölner Geographische Arbeiten.

- (1994a) "Civil and Inner City Disasters," *Erdkunde* 48: 259–74.
- (1994b) "'When the Great Planes Came and Made Ashes of Our City': Towards an Oral History of the Disaster of War," *Antipode* 26: 1–34.
- (1995) *Regions of Risk: Hazards, Vulnerability and Disaster*. London: Longmans.

Higbee, E. (1960) *The Squeeze: Cities without Space*. New York: Morrow.

Higgins, A. (1997) "China Damns Antiquity," *Guardian Weekly* 6 April, 22.

Hildebrand, D. (1997) "Between a Rock and a Hard Place," *Friends of the Innu Newsletter* 1:1–5.

Hindmarsh, R. et al. (1988) *Papers on Assessing the Social Impacts of Development*. Brisbane: Griffith University Institute of Applied Environmental Research.

Hitchens, C. (1992) "Struggle of the Kurds," *National Geographic* 182: 32–61.

Hobsbawm, E. (1991) "Exile: A Keynote Address. Introduction," 65–8 in A. Mack (ed.), *Home: A Place in the World*. Special Edition of *Social Research*. New York: New School for Social Research.

Hodgson, M. (1976) *The Squire of Kootenay West. A Biography of Bert Herridge*. Saanichton, B.C.: Hancock House Publications.

Hollander, J. (1991) "It All Depends," 31–50 in A. Mack (ed.), *Home: A Place in the World*. Special Edition of *Social Research*. New York: New School for Social Research.

Hollsteiner, M. (1977) "The Case of "The People Versus Mr. Urbano Planner y Administrador'," 307–20 in J. Abu-Lughod and R. Hay Jr. (eds.) *Third World Urbanization*. Chicago: Maaroufa.

Holt, W. (1966) *I Haven't Unpacked*. London: Michael Joseph.

Holtby, W. (1936) *South Riding*. London: Collins.

Hong, E. (1996) "Dam Will Wash Away a Culture," *Victoria Times-Colonist* 5 March.

Horowitz, J., and J. Tognoli. (1982) "Role of Home in Adult Development: Women and Men Living Alone Describe Their Residential Histories," *Family Relations* 31: 335–41.

Hoskins, W. (1976) *The Age of Power: The England of Henry VIII 1500–1547*. London: Longmans.

Hudson, R. (1989) *Wrecking a Region: State Policies, Party Politics and Regional Change in North East England*. London: Pion.

Hughes, P. (1988) *V.S. Naipaul*. London: Routledge.

Hume, M. (1993) "Indians Suffered 'Unspeakable Acts'," *The Vancouver Sun* 16 December.

Huszar, L. (1970) "Resettlement Planning," 148–63 in R. Chambers (ed.), *The Volta Resettlement Experience*. New York: Praeger Publishers.

Huxley, A. (1932) *Brave New World*. London: Faber.

Hyman, E. and B. Stiftel (1988) *Combining Facts and Values in Environmental Impact Assessment*. Boulder, Colo.: Westview Press.

IAPS Conference (1992) *Socio-Environmental Metamorphoses*. Proceedings of IAPS Conference, Thessaloniki.

Ignatenko, L. (1998) "Half-Lives: Chernobyl Revisited," *Harper's* May: 17–18.

Illich, I. (1985) *H₂O and the Waters of Forgetfulness: Reflections on the History of "Stuff."* Dallas: Dallas Institute of Humanities and Culture.

Independent Pictures Production. (1962) G. Pinsent, *John and the Missus*.

Ingham, A. (1980) "Using the Space," 166–77 in N. Wates and C. Wolmar (eds.) (1980) *Squatting, the Real Story*. London: Blackrose Press.

Instituto del Tercer Mundo (1997) *The World Guide 1997–8*. Montevideo, Uruguay: I.T.C.

International Rivers Network (2001) *Beyond Big Dams – an NGO Guide to the WCD*. www.irn.org/wcd/bakun.shtml (Accessed 21 February 2001)

Inter Pares. (1994) *Annual Report*. Ottawa: Inter Pares.

Irwin, A. (1992) "Place Attachment and Interpersonal Relationships," 288 in IAPS Conference *Socio-Environmental Metamorphoses*. Proceedings of IAPS Conference, Thessaloniki.

Isaacs, A. (1983) *Without Honor: Defeat in Vietnam and Cambodia*. Baltimore: The Johns Hopkins University Press.

Jackson, I. (1993) *A Century of Dishonor: A Sketch of the United States Government's Dealings with some of the Indian Tribes*. New York: Indian Head Books.

Jackson, J. (1952) "Human, All Too Human Geography," *Landscape* 3: 2–7.

Jackson, K. (1985) *Crabgrass Frontier: The Suburbanization of the United States*. New York: Oxford University Press.

Jacobs, F. (1968) "At the Bottom of the Lake," *Canadian Cattlemen* March: 11, 58.

Jacobs, J. (1961) *The Death and Life of Great American Cities*. New York: Vintage.

James, P. (1989) *Innocent Blood*. London: Penguin.

Janowitz, A. (1990) *England's Ruins: Poetic Purpose and the National Landscape*. Cambridge, Mass.: Blackwell.

Jensen, D. (1996) "A Comparison of the Domicidal Experience of Two Communities in British Columbia," Unpublished paper, Department of Geography, University of Victoria, B.C.

Jhabvala, R. (1981) *Get Ready for Battle*. London: Penguin.

Johnson, N. (1982) *You Can Go Home Again*. Garden City, N.Y.: Doubleday.

Johnson, W. (1984) "Citizen Participation in Local Planning in the U.K. and the U.S.A.: A Comparative Study," *Progress in Planning* 21: 149–221.

Jones, M., and V. Olsen. (1977) *Ilsvikøra, Footdee: To Samfunn, Samme Debatt*. Trondheim, Norway: Galleri Hornemann.

Journal of Environmental Psychology. (1990) "Psychological Fallout from the Chernobyl Nuclear Accident," Special Issue of *Journal of Environmental Psychology*. London: Academic Press.

Jun, J. (1994) "The Assessment of the Three Gorges Project Should Have Involved Sociologists and Anthropologists," 259–61 in Dai Qing (1994) *Yangtze! Yangtze!* London: Earthscan Publications.

Kahneman, D. and J. Knetsch (1992) "Valuing Public Goods: The Purchase of Moral Satisfaction," *Environmental Economics and Management* 22: 57–70.

Kalitsi, E. (1970) "The Organization of Resettlement" 35–57 in R. Chambers (ed.), *The Volta Resettlement Experience*. New York: Praeger Publishers.

Karjalainen, P. (1993) "House, Home and the Place of Dwelling," *Scandinavian Housing and Planning Research* 10: 65–74.

Kastenbaum, R., and R. Aisenberg (1976) *The Psychology of Death: Concise Edition*. New York: Springer.

Kayano, S. (1994) *Our Land was a Forest: An Ainu Memoir*. Boulder, Colo.: Westview.

Kazantzakis, N. (1974) *The Fratricides*. London: Faber.

Kearns, R., and C. Smith. (1994) "Housing, Homelessness, and Mental Health: Mapping an Agenda for Geographical Inquiry," *Professional Geographer* 46: 418–24.

Kent, S. (1992) "Ethnoarchaeology and the Concept of Home: A Cross-Cultural Analysis," 1–11 in *The Ancient Home and the Modern Internationalized Home: Dwelling in Scandinavia*. Trondheim, Norway: University of Trondheim Division of Architectural Design, Norwegian Institute of Technology.

Key, F. (1814) "The Star-Spangled Banner," stanza 2, 436 in J. Bartlett (1955) *Familiar Quotations*. Thirteenth Edition. Boston, Mass.: Little, Brown.

Khalidi, W. (1992) *All That Remains: The Palestinian Villages Occupied and Depopulated by Israel in 1948*. Washington, D.C.: Institute for Palestine Studies.

King, D. (1979) *The Cherokee Indian Nation*. Knoxville, Tenn.: University of Tennessee Press.

King, D., and E. Evans. (1978) "The Trail of Tears: Primary Documents of the Cherokee Removal," *Journal of Cherokee Studies* 3: 129–90.

Kliot, N. (1983) "Dualism and Landscape Tranformation in Northern Sinai – Some Outcomes of the Egypt-Israel Peace Treaty," 173–86 in N. Kliot and S. Wateman (eds.), *Pluralism and Political Geography*. New York: St. Martin's Press.

Knetsch, J. (1983) *Property Rights and Compensation*. Toronto: Butterworths.

Kong, L. (1995) "Music and Cultural Politics: Ideology and Resistance in Singapore," *Transactions, Institute of British Geographers* NS 20: 447–59.

Konrad, G. (1987) *The City Builder*. New York: Penguin.

Kootenay Symposium. (1993 *a*) Notes taken by S. Smith. Kaslo, 27 May.

– (1993 *b*) Notes taken by S. Smith. Nakusp, 2 June.

– (1993 *c*) Notes taken by S. Smith. Cranbrook, 4 June.

– (1993 *d*) Notes taken by S. Smith. Castlegar, 19 June.

Korosec-Serfaty, P. (1984) "The Home from Attic to Cellar," *Journal of Environmental Psychology* 4: 303–21.

Korsching, P., et al. (1980) "Perception of Property Settlement Payments and Replacement Housing among Displaced Persons," *Human Organization* 39: 332–8.

Koyl, M. (1992) *Cultural Chasm: A 1960s Hydro Development and the Tsay Keh Dene Native Community of Northern British Columbia*. Ph.D. Thesis, Department of History, University of British Columbia.

Krawetz, N. (1991) *Social Impact Assessment: An Introductory Handbook*. Halifax, N.S.: Dalhousie University EMDI Environmental Reports, 9.

Kübler-Ross, E. (1992) *On Death and Dying*. New York: Quality Paperback Book Club.

Kyriakos, M. (1994) "What's Home Is Where the Mind Is," *Victoria Times-Colonist* 9 September.

Ladd, F. (1977) "You Can Go Home Again," *Landscape* 21: 15–20.

Lang, R. (1985) "The Dwelling Door: Towards a Phenomenology of Transition," 201–13 in D. Seamon and R. Mugerauer (eds.), *Dwelling, Place and Environment*. New York: Columbia University Press.

Lang, R., and A. Armour. (1981) *The Assessment and Review of Social Impacts*. Ottawa: Federal Environmental Review Office.

Langdon, P. (1994) *A Better Place to Live: Reshaping the American Suburb*. Amherst: University of Massachusetts.

Langford, G. (1957) *Alias O. Henry*. New York: McMillan.

Lapham, L. (1989) "Notebook: Walter Karp (1934–1989)," *Harper's* October: 8–12.

Lasdun, S. (1991) *The English Park: Royal, Private and Public*. London: Andre Deutsch.

Lattey, C. (1980) *Peace River Site C Hydroelectric Development. Social Assessment Update*. Vancouver: Christine Lattey and Associates.

Laurence, M. (1989) *A Bird in the House*. Toronto: McClelland and Stewart.

Lavallée, C., and A. Routhier (1880) "O Canada." First played at a banquet for skaters in Quebec City, 24 June.

Lawrence, D. (1992) "Planning and Environmental Impact Assessment: Never the Twain Shall Meet?," *Plan Canada* July: 22–26.

Lederer, W. and E. Burdick. (1958) *The Ugly American*. New York: Fawcett Crest.

Lelyveld, J. (1985) "Forced Busing in South Africa," *Granta* 17:105–24.

Lenin, V. (1977) "Imperialism, the Highest Stage of Capitalism," 29–35 in J. Abu-Lughod and R. Hay Jr. (eds.) *Third World Urbanization*. Chicago: Maaroufa.

Levi, P. (1989) "My House," *New York Review of Books* 19 January, 25.

Lewin, K. (1946) "Action Research and Minority Problems," *Journal of Social Issues* 1: 34–6.

Lewis, D. (1991) "Drowning by Numbers," *Geographical Magazine* 9: 34–8.

Ley, D. (1977) "Social Geography and the Taken-for-Granted World," *Transactions. Institute of British Geographers* NS 2: 498–512.

Lichfield, N. (1996) *Community Impact Evaluation*. London: University College London Press.

Lieberman, M. (1983) *The Experience of Old Age, Stress, Coping and Survival*. New York: Basic Books.

Lincoln, Y. and E. Guba. (1985) *Naturalistic Inquiry*. Beverly Hills, Cal.: Sage.

Lipsky, M. (1970) *Protest in City Politics*. Chicago: Rand McNally.

Lively, P. (1991) *City of the Mind*. London: Andre Deutsch.

Lloyd, R. et al. (1996) "Basic-Level Geographic Categories," *Professional Geographer* 48: 181–94.

Loewy, R., and W. Snaith (1967) *The Motivations Towards Homes and Housing*. New York: Project Home Committee.

Low, S. (1992) "Cultural Aspects of Place," 286 in *Socio-Environmental Metamorphoses*. Proceedings of IAPS Conference, Thessaloniki.

Lowenthal, D. (1975) "Past Time, Present Place: Landscape and Memory," *The Geographical Review* 65: 1–36.

Lowry, R. (1973) *A.H. Maslow*. Monterey, Cal.: Brooks/Cole.

Lucas, J. (1988) "Places and Dwellings: Wordsworth, Clare and the Anti-Picturesque," 83–97 in D. Cosgrove and S. Daniel (eds.), *The Iconography of Landscape*. Cambridge: Cambridge University Press.

Lupo, A., et al. (1971) *Rites of Way: Transportation in Boston and the U.S. City*. Boston: Little, Brown.

MacAskill, E. (2000) "Evicted islanders win right of return," *Guardian Weekly* 15 November.

– (2001) "Building unbearable lives," *Guardian Weekly* 18 January.

Mack, A. (ed.) (1991) "Home: A Place in the World," Special edition of *Social Research*. New York: New School for Social Research.

Mackie, K. (1981) *An Exploration of the Idea of Home in Human Geography*. Toronto: University of Toronto Master of Arts research paper.

Maclean's (1993) "Newfoundland: Can the Province be Saved?" 23 August: 18–30.

Macpherson, C. (1978) *Property: Mainstream and Critical Positions*. Toronto: University of Toronto Press.

Maddrell, P. (1990) *The Bedouin of the Negev*. London: The Minority Rights Group.

Madely, J. (1985) *Diego Garcia: A Contrast to the Falklands*. London: The Minority Rights Group.

Makiya, K. [Samir Al-Khalil] (1994) *Cruelty and Silence: War, Tyranny and Uprising in the Arab World*. London: Penguin.

Malcomson, S. (1994) *Borderlands: Nation and Empire*. Boston: Faber.

Malczewski, J. (1995) Review of *The Right Place* by B. Massam. *Canadian Geographer* 39: 377–8.

Mallaby, S. (1992) *After Apartheid*. New York: Times Books.

Manchester Guardian Weekly (1989) "Old Soviet Tensions Now in the Open," *Manchester Guardian Weekly* 18 June.

– (1992) "Cultural 'Cleansing'," *Manchester Guardian Weekly* 25 October.

Marc, O. (1972) *Psychoanalyse de la Maison*. Paris: Seuil.

– (1977) *Psychology of the House*. London: Thames and Hudson.

Marcuse, H. (1964) *One-Dimensional Man: Studies in the Ideology of Advanced Industrial Society*. Boston: Beacon Press.

Markoutsas, E. (1992) "Changing with the Times: 40 Years of Laura Ashley," *Victoria Times-Colonist* 20 December, C1.

Marnham, P. (1987) *Fantastic Invasion: Dispatches from Africa*. London: Penguin.

Marples, D. (1988) *The Social Impact of the Chernobyl Disaster*. Edmonton: University of Alberta Press.

Marris, P. (1969) "A Report on Urban Renewal in the United States," 113–34 in L. Duhl (ed.) *The Urban Condition*. New York: Simon and Schuster.

– (1974, Revised Edition 1986) *Loss and Change*. New York: Pantheon.

– (1980) "The Uprooting of Meaning," 101–16 in G. Coelho and P. Ahmed (eds.) *Uprooting and Development: Dilemmas of Coping with Modernization*. New York: Plenum.

– (1989) *The Dreams of General Jerusalem*. London: Bloomsbury.

Martin, J. (1984) *Miss Manner's Guide to Rearing Perfect Children*. New York: Penguin.

Martin, R. (1991) *Gerard Manley Hopkins: A Very Private Life*. New York: Putnam's.

Mason, L., and E. Hereniko (1987) *In Search of a Home*. Suva, Fiji: Institute of Pacific Studies, University of the South Pacific.

Massam, B. (1993) *The Right Place: Shared Responsibility and the Location of Public Facilities*. London: Longmans.

– (1995) Review of *LLRW Disposal Facility Siting* by Vari et al. *Canadian Geographer* 39: 376–7.

– (1999) "Past President's Address: Geographical Perspectives on the Public Good," *Canadian Geographer* 43: 346–62.

Matthews, R. (1970) *Communities in Transition: An Examination of Government Initiated Community Migration in Rural Newfoundland*. Ph.D. Thesis, University of Minnesota.

– (1976) *"There's No Better Place Than Here:" Social Change in Three Newfoundland Communities*. Toronto: Peter Martin Associates.

Maxwell, G. (1963) *Ring of Bright Water*. London: Pan Books.

McAllister, D. (1980) *Evaluation in Environmental Planning*. Cambridge, Mass.: M.I.T. Press.

McCallum, H., and K. McCallum (1975) *This Land Is Not for Sale*. Toronto: Anglican Book Centre.

McCandlish, J. (1971) "Kootenay Residents Bitter Over Libby Dam Land Grab," *Vancouver Sun* 18 October.

McCarron, et al. (1994) "Communication, Belonging and Health," 57–72 in M. Hayes et al. (eds.) *The Determinants of Population Health*. Victoria, B.C.: Western Geographical Series.

McDaniels, T. (1993) "Contingent Valuation and Multiple Objective Approaches Compared," in Province of British Columbia. Ministry of Environment, Lands and Parks. *Full Cost Accounting & The Environment*. Seminar Proceedings (19 March 1993) Victoria, B.C.: Ministry of Environment, Lands and Parks.

McDonald, M., and J. Muldowny (1982) *TVA and the Dispossessed: Resettlement of Population in the Norris Dam Area*. Knoxville: University of Tennessee Press.

McDowell, L., and D. Massey (1984) "A Women's Place," 128–47 in O. Massey and J. Allen (eds.), *Geography Matters*. Cambridge, Mass.: Cambridge University Press.

McKie, D. (1973) *A Sadly Mismanaged Affair*. London: Croom Helm.

Meir, A. (1997) *As Nomadism Ends: The Israeli Bedouin of the Negev*. Boulder, Colo.: Westview.

Meredith, M. (1979) *The Past Is Another Country*. London: Andre Deutsch.

Mickleburgh, R. (1997) "China Diverts Mighty Yangtze to Cheers," *Toronto Globe and Mail* 10 December, A8.

– (1998) "Corruption Drilling Holes in China's Huge Dam Project," *Toronto Globe and Mail* 13 March, A9.

Milbraith, L. (1965) *Political Participation*. Chicago: Rand McNally.

Miller, J. (1989) *Skyscrapers Hide the Heavens: A History of Indian-White Relations in Canada*. Toronto: University of Toronto Press.

Million, L. (1992) *"It Was Home": A Phenomenology of Place and Involuntary Displacement as Illustrated by the Forced Dislocation of Five Southern Alberta Families in the Oldman River Dam Flood Area*. Ph.D. dissertation, Saybrook Institute, San Diego, Cal.

Mills, D. (1991) *Rebirth of the Corporation*. New York: Wiley.

Mintzberg, H. (1993) *The Rise and Fall of Strategic Planning*. New York: The Free Press.

Mishan, E.J. (1986) *Economic Myths and the Mythology of Economics*. Atlantic Highlands, N.J.: Humanities Press International.

Mistry, R. (1991) *Such a Long Journey*. Toronto: McClelland and Stewart.

– (1995) *A Fine Balance*. Toronto: McClelland and Stewart.

Mitchell, L. (1981) *Witnesses to a Vanishing America*. Princeton, N.J.: Princeton University Press.

Moore, T. (1996) *The Re-Enchantment of Everyday Life*. New York: Harper Collins.

Morris, B. (1987) *The Birth of the Palestine Refugee Problem 1947–1949*. Cambridge: Cambridge University Press.

Moses, R. (1970) *Public Works: A Dangerous Trade*. New York: McGraw-Hill.

Moss-Kanter, R. (1983) *The Change Masters*. New York: Simon and Schuster.

Motz, M., and P. Browne (1988) *Making the American Home*. Bowling Green, Ohio: Bowling Green State University Popular Press.

Mui, H. (1992) "Without a Room of One's Own: Woman's Place in Culture and Space," 293 in *Socio-Environmental Metamorphoses*. Proceedings of IAPS Conference, Thessaloniki.

Munif, A. (1987) *Cities of Salt*. New York: Random House.

Murphy, P. (1995) "Official Turns Tables on Royal Commission," *Victoria Times-Colonist* 2 March, A14.

Mumford, L. (1961) *The City in History: Its Origins, Its Transformations and Its Prospects*. New York: Harcourt, Brace & World.

Musil, J. (1972) "Sociology of Urban Redevelopment Areas: A Study from Czechoslovakia," 298–303 in G. Bell and J. Tyrwhitt (eds.) *Human Identity in the Urban Environment*. London: Penguin.

Myers, G. (1998) "Intellectual of Empire: Eric Dutton and Hegemony in British Africa," *Annals of the Association of American Geographers* 88: 1–27.

Myers, N. (ed.) (1984) *Gaia: An Atlas of Planet Management*. New York: Anchor Doubleday.

Nakano, A. (1983) *Broken Canoe: Conversations and Observations in Micronesia*. St. Lucia: University of Queensland Press.

Nation (Bangkok) 23 December 1994.

Neil, C., et al. (1992) *Coping with Closure: An International Comparison of Mine Town Experiences*. London: Routledge.

New Internationalist September 1991.

New Internationalist November 1995.

New Internationalist May 1996.

New Internationalist April 1997.

Newitt, M. (1995) *A History of Mozambique*. London: Hurst.

Nicosia, G. (1990) "Kerouac: Writer Without a Home," 19–39 in P. Anctil et al. (eds.), *Un Homme Grand*. Ottawa: Carleton University Press.

Nordland, R. (1991) "Saddam's Secret War," *Newsweek* 10 June.

Norris, C. (1990) "Stories of Paradise: What Is Home When We Have Left It," *Phenomenology & Pedagogy* 8: 237–44.

Nuttal, M. (1992) *Arctic Homeland. Kinship, Community and Development in Northwest Greenland*. Toronto: University of Toronto Press.

Obermayer, N., and J. Pinto (1994) *Managing Geographic Information Systems*. New York: Guilford.

Oliver-Smith, A. (1991) "Involuntary Resettlement, Resistance and Political Empowerment," *Journal of Refugee Studies*. 4: 132–49.

Orme, A. (1970) *Ireland*. Chicago: Aldine.

Orr, D. (1994) "Professionalism and the Human Prospect," *Conservation Biology* 8: 9–11.

Orwell, G. (1945) *Animal Farm*. London: Secker and Warburg.

– (1949) *1984*. London: Secker and Warburg.

Pallister, D. (2000) "Islanders Sue US over Impact of Rio Tinto Mine," *Guardian Weekly* 14 September.

Pandey, B. (1998) *Depriving the Underprivileged for Development*. Bhubaneswar, Orissa, India: Institute for Socio-Economic Development.

Parmenter, B. (1994) *Giving Voice to Stones; Place and Identity in Palestinian Literature*. Austin: University of Texas Press.

Partridge, W. (1989) "Involuntary Resettlement in Development Projects," *Journal of Refugee Studies* 2: 373–84.

Pascal, G. (1997) "Nonviolence and the People of the First Nations," *Ploughshares Monitor* 18: 17–18.

Payne, J. (1823) "Clari, the Maid of Milan," line from opera, 464 in J. Bartlett (1955) *Familiar Quotations*. Thirteenth Edition. Boston, Mass.: Little, Brown.

Pearce, H. (1991) "The Flooding of a Nation," *Geographical Magazine* November: 18–21.

Peled, A., and O. Ayalon (1988) "The Role of the Spatial Organization of the Home in Family Therapy: A Case Study," *Journal of Environmental Psychology* 8: 87–106.

Pennartz, P. (1986) "Atmosphere at Home: A Qualitative Approach," *Journal of Environmental Psychology* 6: 135–53.

Pepper, D. (1980) "Environmentalism, the 'Lifeboat Ethic' and Anti-Airport Protest," *Area* 12: 177–82.

Perlman, J. (1982) "Favela Removal: The Eradication of a Lifestyle," 225–43 in A. Hansen and A. Oliver-Smith (eds.), *Involuntary Migration and Resettlement*. Boulder, Col.: Westview Press.

Perman, D. (1973) *Cublington: A Blueprint for Resistance*. London: The Bodley Head.

Peters, T. (1987) *Thriving on Chaos: Handbook for a Management Revolution*. New York: Knopf.

Phatkul, N. (1994) "Resettlement Planned for Wildlife Sanctuary Dwellers" and "Protesters, Human Rights Panel Meet over Said Police Brutality," *The Nation* (Bangkok) 23 December, A3.

Pilger, J. (1998) *Hidden Agendas*. London: Vintage.

Platzky, L., and C. Walker (1985) *The Surplus People, Forced Removals in South Africa*. Johannesburg: Ravan Press.

Plummer, K. (1983) *Documents of Life*. London: Allen and Unwin.

Pocock, D. (1980) "Place and the Novelist," *Transactions, Institute of British Geographers* NS 6: 337–47.

Pollon, E., and S. Matheson. (1989) *This Was Our Valley*. Calgary, Alberta: Detselig Enterprises.

Porteous, J. (1972) "Urban Transplantation in Chile," *The Geographical Review* 62: 455–78.

– (1975) "Quality of Life in B.C. Company Towns," *Contact* 7: 26–37.

– (1976) "Home: The Territorial Core," *The Geographical Review* 66: 383–90.

– (1977) *Degrees of Latitude*. Saturna Island BC: Saturnalia.

– (1977) *Environment and Behavior: Planning and Everyday Urban Life*. Reading, Mass.: Addison-Wesley.

– (1981) *The Modernization of Easter Island*. Victoria, B.C.: Western Geographical Series.

– (1988) "Topocide," 75–93 in J. Eyles and D. Smith (eds.), *Qualitative Methods in Geography*. Cambridge: Polity Press.

– (1989) *Planned to Death: The Annihilation of a Place Called Howdendyke*. Manchester: Manchester University Press.

– (1990) *Landscapes of the Mind*. Toronto: University of Toronto Press.

– (1992a) "Domicide: The Destruction of Home," A Keynote Address for the Symposium "The Ancient Home and the Modern Internationalised Home: Dwelling in Scandinavia." Trondheim, Norway: University of Trondheim Division of Architectural Design, Norwegian Institute of Technology.

– (1992b) "The Mutual Impenetrability of Worlds of Discourse," *Environmental and Architectural Phenomenology Newsletter* 3: 10–11.

– (1995) "Planning Applications as Evidence Against the Rich and Powerful," *Area* 27: 137–9.

– (1996) *Environmental Aesthetics: Ideas, Politics and Planning*. London: Routledge.

Portugali, J. (1989) "Nomad Labour: Theory and Practice in the Israel-Palestinian Case," *Transactions, Institute of British Geographers* NS 14: 207–20.

Powell, W. (1994) "Facing the Floodgates," *Sunday Telegraph* 10 July.

Prebble, J. (1969) *The Highland Clearances*. London: Penguin.

Probe International (1992) "World Bank Continues to Support Controversial Sardar Sarovar Dam in India, Displacing 240,000 People," *Probe Alert* December.

– (1993a) "Spotlight on China," *Probe Alert* September.

– (1993b) "Our Record of Accomplishment," in mail-out from Probe International, 26 November.

– (1994) "Dammed" *Probe International Update*.

– (1995) *World Bank Backgrounder* #52. www.probeinternational.org (accessed 21 February 2001).

Proshansky, H., et al. (1983) "Place Identity: Physical World Socialization of the Self," *Journal of Environmental Psychology* 3: 57–83.

Ptolemy (1994) "Helping Murphy's Law," *Association of American Geographers Newsletter* 29: 20.

Ragon, M. (1981) *The Space of Death*. Charlottesville, Va.: University Press of Virginia.

Rainwater, L. (1966) "Fear and the House-as-Haven in the Lower Class," *Journal of the American Institute of Planners* 32: 25–31.

Rakoff, R. (1977) "Ideology in Everyday Life: The Meaning of the House," *Politics and Society* 7: 85–104.

Randall, J., and R. Ironside. (1996) "Communities on the Edge: An Economic Geography of Resource-Dependent Communities in Canada," *Canadian Geographer* 40: 17–35.

Ransom, D. (ed.) (1996) "Homelessness," *New Internationalist* September, whole issue.

Rao, K., and C. Geisler (1988) "The Social Consequences of Protected Areas Development on Resident Populations" in R. Hindmarsh et al. (eds.), *Papers on Assessing the Social Impacts of Development*. Brisbane: Griffith University Institute of Applied Environmental Research.

Rapoport, A. (1969) *House Form and Culture*. Englewood Cliffs, N.J.: Prentice-Hall.

– (1992) "A Critical View of the Concept of Home," 1–39 in *The Ancient Home and the Modern Internationalized Home: Dwelling in Scandinavia*. Trondheim, Norway: University of Trondheim Division of Architectural Design, Norwegian Institute of Technology.

Raskin, M. (1986) *The Common Good*. New York: Routledge and Kegan Paul.

Raunet, D. (1984) *Without Surrender, Without Consent*. Toronto: Douglas and McIntyre.

Reed, G. (1979) "Postremoval Factionalization in the Cherokee Nation," 148–63 in D. King (ed.) *The Cherokee Indian Nation*. Knoxville, Tenn.: University of Tennessee Press.

Reed, M. (1984) *Citizen Participation and Public Hearings: Evaluating Northern Experiences*. Victoria, B.C.: Department of Geography, University of Victoria.

Rees, R. (1982) "In a Strange Land ... Homesick Pioneers on the Canadian Prairie," *Landscape* 26: 1–9.

Relph, E. (1970) "An Enquiry into the Relations between Phenomenology and Geography," *The Canadian Geographer* 14: 193–201.

– (1976) *Place and Placelessness*. London: Pion.

Restak, J. (1979) *The Brain*. New York: Fawcett.

Richburg, K. (1995) "Dictators Flourish in Africa. Western Aid No Longer Tied to Reform," *International Herald Tribune* 3 January.

Riley, R. (1992) "Place Attachment: A Conceptual Exploration." Paper presented at the IAPS 12th International Conference, Marmaras, Greece.

Ripley, J. (1964) "The Columbia River Scandal," *Engineering Contract and Record* 4: 45–60.

Road Alert! (1998) *Top Tips for Wrecking Roadbuilding*. London: Road Alert!

Roberts, T. (Focus Consultants of Victoria, B.C.) (1994) Discussion regarding the use of "victim impact statements."

Robertson, M. (1993) "Kemano II – The Cheslatta Reserves Surrender," *Project North B.C.* 4: 1, 5, 10–11.

Rogerson, C., and S. Parnell (1989) "Fostered by the Laager: Apartheid Human Geography in the 1980s," *Area* 21: 13–26.

Rohe, W., and S. Mouw (1991) "The Politics of Relocation," *APA Journal* 57: 57–68.

Rogge, J. (1987) *Refugees: A Third World Dilemma*. New York: Rowman and Littlefield.

Room, A. (1985) *Dictionary of Confusing Words and Meanings*. London: Routledge.

Rossi, P., et al. (1982) *Natural Hazards and Public Choice*. New York: Academic Press.

Routledge, P. (1993) *Terrains of Resistance: Nonviolent Social Movements and the Contestation of Place in India*. Westport, Conn.: Praeger.

– (1997) "The Imagineering of Resistance: Pollock Free State and the Practice of Postmodern Politics," *Transactions, Institute of British Geographers* NS 22: 359–76.

Rowe, S. (1990) *Home Place: Essays on Ecology*. Edmonton: Newest.

Rowles, G. (1978) *Prisoners of Space?* Boulder, Colo.: Westview Press.

Rudnicki, W., and H. Dyck (1986) "The Government of Aboriginal Peoples in Other Countries," 378–91 in R. Ponting (ed.) *Arduous Journey*. Toronto: McClelland and Stewart.

Ruitenbeek, H., and C. Cartier (1993) "A Critical Perspective on the Evaluation of the Narmada Projects from the Discipline of Ecological Economics," Paper presented to The Narmada Forum: Workshop on the Narmada Sagar & Sardar Sarovar. New Delhi, India: draft paper.

Rushdie, S. (1988) *The Satanic Verses*. Dover, Del.: The Consortium.

Ryan, A. (1987) *Property*. Milton Keynes: Open University Press.

Ryan, E. (1969) "Personal Identity in an Urban Slum," 135–50 in L. Duhl (ed.) *The Urban Condition*. New York: Simon and Schuster.

Rybczynski, W. (1986) *Home*. New York: Viking Press.

– (1989) *The Most Beautiful House*. New York: Viking Press.

Saddul, P. (ed.) (1996) *Atlas of Mauritius*. Port Louis, Mauritius: Editions de l'Ocean Indien.

Sadler, B. (1989) "National Parks, Wilderness Preservation and Native People in Northern Canada," *Natural Resources Journal* 29: 185–204.

Said, E. (1993) *Culture and Imperialism*. New York: Vintage.

– (1994) *The Politics of Dispossession: The Struggle for Palestinian Self-Determination 1969–1994*. New York: Pantheon.

Saile, D. (1985) "The Ritual Establishment of Home," 87–107 in I. Altman and C. Werner (eds.), *Home Environments*. New York: Plenum Press.

Salasan Associates Inc. (19 June 1993) "Columbia-Kootenay Symposium Saturday Night Report," photocopy.

– (17 August 1993) "Columbia-Kootenay Symposium Summary Report," photocopy.

Sandars, N. (1960) *The Epic of Gilgamesh*. London: Penguin.

Sanford, N. (1970) "Whatever Happened to Action Research?" *Journal of Social Issues* 26: 3–23.

Sayegh, K. (1972) *Canadian Housing: A Reader*. Waterloo, Ont.: University of Waterloo Faculty of Environmental Studies.

Schama, S. (1991) "Homelands," 11–30 in A. Mack (ed.), *Home: A Place in the World*. Special Edition of *Social Research*. New York: New School for Social Research.

Scherbak, I. (1989) *Chernobyl. A Documentary Story*. Basingstoke, Hants: McMillan Press.

Schutz, A. (1971) "The Homecomer," 106–19 in A. Broderson (ed.), *Collected Papers II: Studies in Social Theory*. The Hague: Martinus Nijhoff.

Scudder, T. (1982) *No Place to Go: Effects of Compulsory Relocation on Navajos*. Philadelphia: Institute for the Study of Human Issues.

Scudder, T., and E. Colson. (1982) "From Welfare to Development: A Conceptual Framework for the Analysis of Dislocated People," 267–87 in A. Hansen and A. Oliver-Smith (eds.), *Involuntary Migration and Resettlement*. Boulder, Colo.: Westview.

Seabrook, J. (1988) *The Race to Riches: The Human Cost of Wealth*. Basingstoke, Hampshire: Marshall Pickering.

– (1997) "The City, Our Stepmother," *New Internationalist* May: 7–16.

Seamon, D. (1979) *Geography of the Lifeworld: Movement, Rest and Encounter*. London: Croom Helm.

Segal, W. (1973) "Home Sweet Home," *Royal Institute of British Architects Journal* 10: 477–80.

Séguin, R. (1994) "Quebec Shelves Great Whale Project," *Globe and Mail* 19 November.

Sesser, S. (1994) *The Lands of Charm and Cruelty: Travels in Southeast Asia.* New York: Vintage.

Sharpe, T. (1975) *Blott on the Landscape.* London: Secker and Warburg.

Shaw, S. (1990) "Returning Home," *Phenomenology & Pedagogy* 8: 225–37.

Shelden, M. (1991) *Orwell: The Authorized Biography.* New York: Harper Collins.

– (1994) *Graham Greene: The Man Within.* London: Heinemann.

Shelley, P.B. (1941) "Ozymandias of Egypt," 251 in F. Palgrave (ed.) *The Golden Treasury.* London: Oxford University Press.

Shihab, S. (1997) "Turkmenistan Has Pipedream of a Golden Age," *The Guardian Weekly* 14 September.

Shiva, V. (ed.) (1994) *Close to Home: Women Reconnect Ecology, Health and Development Worldwide.* Philadelphia: New Society.

Short, J. (1991) *Imagined Country: Society, Culture and Environment.* London: Routledge.

Simmie, J. (1974) *Citizens in Conflict: The Sociology of Town Planning.* London: Hutchinson.

Simpson, J. (ed.) (1995) *The Oxford Book of Exile.* New York: Oxford University Press.

Simpson-Housley, P., and A. de Man (1987) *The Psychology of Geographical Disasters.* North York, Ont.: York University Geographical Monographs no. 18.

Sixsmith, J. (1986) "The Meaning of Home: An Exploratory Study of Environmental Experience," *Journal of Environmental Psychology* 6: 281–8.

Slim, H., and P. Thompson (1995) *Listening for a Change: Oral Testimony and Community Development.* Philadelphia: New Society.

Smith, D. (1978) "Involuntary Population Movement in South Africa," *Area* 10: 87–8.

– (1987) *Apartheid in South Africa.* Cambridge: Cambridge University Press.

Smith, N. (1996) *The New Urban Frontier: Gentrification and the Revanchist City.* New York: Routledge.

Smith, P. (1992) Personal Communication quoted in J. Porteous "Domicide: The Destruction of Home," A Keynote Address for the symposium The Ancient Home and the Modern Internationalised Home: Dwelling in Scandinavia. Trondheim, Norway, 1992.

Soderstrom, E. (1981) *Social Impact Assessment. Experimental Methods and Approaches.* New York: Praeger.

Somerville, P. (1992) "Homelessness and the Meaning of Home: Rooflessness or Rootlessness?" *International Journal of Urban and Regional Research* 16: 529–39.

Sommer, R. (1997) "Utilization Issues in Environment – Behavior Research," 347–68 in G. Moore and R. Marans (eds.) *Advances in Environment, Behavior and Design*, volume 4. New York: Plenum.

Sommer, R., and B. Sommer (1984) *Scenic Drowning*. Davis, Cal.: Greycats Press.

Sopher, D. (1979) "The Landscape of Home: Myth, Experience and Social Meaning," 129–49 in D. Meinig (ed.), *Interpretations of Ordinary Landscapes: Geographical Essays*. Oxford: Oxford University Press.

Spiegel, H., and C. Mittenthal (1968) "The Many Faces of Citizen Participation," 28–39 in H. Spiegel, (ed.), *Citizen Participation in Urban Development*. Washington, D.C.: National Institute for Applied Behavioural Science.

Stackhouse, J. (1998) "Decline and Fall of the Big-Dam Era," *Globe and Mail* 28 March, D4.

Stanley, P. (1993) "Banaba: The Story of an Island," *Tok Blong SPPF* (Victoria, B.C.) 44: 25–6.

Steed, G. (1988) "Geography, Social Science, and Public Policy: Regeneration through Interpretation," *The Canadian Geographer* 32: 2–13.

Steele, J. (1995) "Ash Wednesday: The Night It Rained Fire," *The Guardian* 9 February.

Stinson, A. (ed.) (1974) *Citizen Action: An Annotated Bibliography of Canadian Case Studies*. Ottawa: Carleton University Centre for Social Welfare.

Stoett, P. (1999) *Human and Global Security: An Exploration of Terms*. Toronto: University of Toronto Press.

Strauss, E. (1954) *Sir William Petty: Portrait of a Genius*. London: The Bodley Head.

Sullivan, L. (1995) "The Three Gorges: Dammed If They Do?" *Current History* 94: 266–9.

Sutton, K., and R. Lawless. (1978) "Population Regrouping in Algeria: Traumatic Change and the Rural Settlement Pattern," *Transactions, Institute of British Geography* NS 3: 331–50.

Swainson, N. (1979) *Conflict Over the Columbia: The Canadian Background to an Historic Treaty*. Montreal: McGill-Queen's University Press.

– (1995) Discussion with S. Smith, 27 February.

Sweeney, J. (1998) "Press," *The Oldie* 110, May.

Swift, R. (1993) "Myth and Memory," *New Internationalist* September: 4–7.

– (1995) "Flood of Protest," *New Internationalist* November: 7–16.

Sylvester, C. (1991) *Zimbabwe. The Terrain of Contradictory Development*. Boulder, Colo.: Westview.

Talbott, J. (1980) *The War Without a Name. France in Algeria, 1954–1962*. New York: Knopf.

Taylor, M. (1987) *Cultural Components and the Decision to Resettle the Palestinian Refugees*. Victoria, B.C.: Honours Thesis, Department of Geography, University of Victoria.

Taylor, R., and S. Brower (1985) "Home and Near Home Territories," 183–212 in L Altman and C. Werner (eds.), *Home Environments*. New York: Plenum Press.

Terkenelli, T. (1994) "The Idea of Home: A Cross-Cultural Perspective," *Environmental and Architectural Phenomenology Newsletter* 5: 4.

Tester, F., and P. Kulchyski. (1994) *Tammarnitt (Mistakes): Inuit Relocation in the Eastern Arctic 1939–63*. Vancouver, B.C.: University of British Columbia Press.

Thomas, W. (1995) *Scorched Earth: The Military's Assault on the Environment*. Philadelphia: New Society.

Tindall, G. (1991) *Countries of the Mind*. London: Hogarth Press.

Tinder, G. (1980) *Community: Reflections on a Tragic Ideal*. Baton Rouge: Louisiana State Press.

Todd, E. (1992) *The Law of Expropriation and Compensation in Canada*. Scarborough, Ont.: Carswell.

Tognoli, J. (1980) "Differences in Women's and Men's Responses to Domestic Space," *Sex Roles* 6: 833–42.

– (1987) "Residential Environments," 655–90 in D. Stokols and I. Altman (eds.), *Handbook of Environmental Psychology*. New York: Wiley Interscience.

Trimble, J. (1978) "Issues of Forced Relocation and Migration of Cultural Groups," in *Society for Intercultural Education, Training and Research, Overview of Intercultural Education, Training and Research Areas*. Washington, D.C.: Georgetown University.

– (1980) "Forced Migration: Its Impact on Shaping Coping Strategies," 449–78 in G. Coelho and P. Ahmed (eds.), *Uprooting and Development: Dilemmas of Coping with Modernization*. New York: Plenum.

Tripp, M. (1998) *The Emergence of National Parks in Russia*. Victoria, B.C.: Ph.D. Thesis, Department of Geography, University of Victoria.

Tuan, Y-F (1961) "Topophilia–or Sudden Encounter with Landscape," *Landscape* 11: 29–32.

– (1971) "Geography, Phenomenology, and the Study of Human Nature," *Canadian Geographer* 15: 188–92.

– (1974) *Topophilia*. Englewood Cliffs, N.J.: Prentice Hall.

– (1975a) "Place, an Experiential Perspective," *Geographical Review* 65: 151–65.

– (1975b) "Home as an Environmental and Psychological Concept," *Landscape* 20: 2–9.

– (1977) *Space and Place: The Perspective of Experience*. Minneapolis, Minn.: University of Minnesota Press.

– (1980) "Rootedness versus Sense of Place," *Landscape* 24: 3–8.

– (1993) *Passing Strange and Wonderful*. Washington, D.C.: Island Press.

Turnbull, C. (1972) *The Mountain People*. New York: Simon and Schuster.

Turner, F. (1974) *North American Indian Reader*. New York: Viking.

Turner, J. (1984) *The Architect as Enabler of User House Planning and Design*. Stuttgart: Karl Dramer Verlag.

Twigger-Ross, C., and D. Uzzell. (1996) "Place and Identity Processes," *Journal of Environmental Psychology* 16: 205–21.

Updike, J. (1978) *The Coup*. New York: Fawcett.

Uris, L. (1976) *Trinity: A Novel of Ireland*. New York: Bantam Doubleday.

Urquhart, J. (1997) *The Underpainter*. Toronto: McClelland and Stewart.

Van der Gaag, J. (1996) "A Refugee Settlement," *New Internationalist* September: 7–19.

Vari, A., et al. (1994) *LLRW Disposal Facility Siting: Success and Failure in Six Countries*. Dordrecht: Kluwer.

Vassilikos, V. (1991) Z. New York: Four Walls, Eight Windows.

Vidal, J. (1997) "A Tribe's Suicide Pact," *Guardian Weekly* 12 October.

– (1998) "Woman Power Halts Work on Indian Dam," *Guardian Weekly* 18 January.

Vidich, A. and J. Bensman (1960) *Small Town in Mass Society*. Princeton, N.J.: Princeton University Press.

Violich, F. (1993) "Dalmatia, Urban Identity and the War, 1991–1993: Seeking Meaning in Urban Places, " *Environmental and Architectural Phenomenology Newsletter* 4: 11–13.

Vizinczey, S. (1986) "Engineers of a Sham: How Literature Lies about Power," *Harper's* 272: 69–73.

Wadden, M. (1996) *Nitassinan: The Innu Struggle to Reclaim their Homeland*. Vancouver: Douglas and McIntyre.

Wagg, D. (1993) "Kemano 1 Washed Away 50 Graves," *The Watershed* 1: 1–2.

Walker, R. (1985) *Applied Qualitative Research*. London: Gower.

Walklate, S. (1989) *Victimology: The Victim and the Criminal Justice Process*. London: Unwin Hyman.

Wallace, D. (1968) "The Conceptualizing of Urban Renewal," *University of Toronto Law Journal* 18: 248–58.

Walter, A. (1990) *Funerals: And How to Improve Them*. London: Hodder and Stoughton.

Ward, M. (1993) "The Power of Power Smart: B.C. Hydro's Bright Idea," *Canadian Environment Business Magazine* 3: 8–11.

Washington Post (1993) "The Seeds of Hatred," *Guardian Weekly* 10 January, 146.

Waterfield, D. (1970) *Continental Waterboy*. Vancouver: Clarke, Irwin.

Waterson, B. (1989) "Calvin and Hobbes," *Victoria Times-Colonist* 9 May.

Wates, N., and C. Wolmar (eds.) (1980) *Squatting, the Real Story*. London: Blackrose Press.

Watson, S., and H. Austerberry (1986) *Housing and Homelessness: A Feminist Perspective*. London: Routledge and Kegan Paul.

Watts, I. (1930) "O God, Our Help in Ages Past," Psalm 662 in *The Hymnary of the United Church of Canada*. Toronto: The United Church Publishing House.

Weil, S. (1952) *The Need for Roots*. New York: Putnam.

Weiler, J. (1980) *Guidelines on the Man-Made Heritage Component of Environmental Assessments*. Toronto: Ontario Minister of Culture and Recreation.

Weisner, T. and J. Weibel (1981) "Home Environments and Family Lifestyles in California," *Environment and Behaviour* 13: 416–60.

Weisstub, D. (1986) "Epilogue: On the Rights of Victims" 317–22 in Fallah, E. (ed.), *From Crime Policy to Victim Policy*. Houndmills, Hampshire: Macmillan.

Wellard, J. (1973) *By the Waters of Babylon*. Newton Abbot U.K.: Readers Union.

Welsh, A. (1971) *The City of Dickens*. Oxford: Clarendon Press.

Werner, C. et al. (1985) "Temporal Aspects of Homes: A Transactional Perspective, " 1–32 in I. Altman and C. Werner (eds.), *Home Environments*. New York: Plenum Press.

– (1989) "Inferences about Homeowners' Sociability: Impact of Christmas Decorations and Other Cues," *Journal of Environmental Psychology* 9: 279–96.

Western, J. (1981) *Outcast Cape Town*. Minneapolis: University of Minnesota Press.

– (1992) *A Passage to England: Barbadian Londoners Speak of Home*. London: UCL Press.

White, G. (ed.) (1961) *Papers on Flood Problems*. Chicago: University of Chicago Department of Geography Research Paper No. 29.

White, L. (1987) *Creating Opportunities for Change*. Boulder, Colo.: Lynne Rienner Publishers.

Whyte, W. (ed.) (1991) *Participatory Action Research*. Newbury Park, Cal.: Sage.

Wiggins, M. (1989) *John Dollar, A Novel*. New York: Harper and Row.

Wikström, T. (1994) *Between the Home and the World: Space and Housing Interaction in Housing Areas of the Forties and Fifties*. Ph.D. Dissertation (Swedish with English summary). Department of Building Functions and Analysis, Lund, Sweden.

Wilkes, K. (1992) "Lead Upon Gold," *Oxford Today* 5: 22–5.

Will, G. (1986) "A Man's Curtilage Is Not His Castle," *The Washington Post* 15 June.

Williams, P. (1993) "The Promise and the Glory," *Equinox* 12: 83–94.

Williams, R. (1975) *The Country and the City*. St Albans, UK: Paladin.

Willmott, P., and M. Young (1957) *Family and Kinship in East London*. London: Routledge and Kegan Paul.

Wilson, F. (1991) Book Review of *Scholarship Reconsidered: Priorities of the Professoriate* (A Special Report for the Carnegie Foundation for the Advancement of Teaching) by E.L. Boyer. *CAUT Bulletin ACPU* March: 25.

Wilson, J. (1973) *People in the Way; The Human Aspects of the Columbia River Project*. Toronto: University of Toronto Press.

– (1993) Discussion at Kootenay Symposium and follow-up correspondence, 27 June.

Wilson, J. and M. Conn (1983) "On Uprooting and Rerooting: Reflections on the Columbia River Project," *BC Studies* 58: 40–54.

Winchester, S. (1985) *Outposts*. London: Hodder and Stoughton.

Winning, A. (1990) "Homesickness," *Phenomenology & Pedagogy* 8: 245–58.

– (1991) "The Speaking of Home, " *Phenomenology & Pedagogy* 9: 180–92.

Wishart, D. (1979) "The Dispossession of the Pawnee," *Annals of the Association of American Geographers* 69: 382–401.

Wisner, B., et al. (1991) "Participatory and Action Research Methods," 271–95 in E. Zube and G. Moore (eds.) *Advances in Environment, Behavior and Design* Volume 3. New York: Plenum.

Wolf, P. (1981) *The Human Side of Environmental Impact Assessment: A Federal Perspective*. Canada: Federal Environmental Assessment Review Office Occasional Paper no. 7.

Wood, W. (1994) "Forced Migration: Local Conflicts and International Dilemmas," *Annals of the Association of American Geographers* 84: 607–34.

Woon, Y-F. (1994) "The Pearl River Delta Region, Winter 1993," *Asia-Pacific News* Newsletter of the Centre for Asia Pacific Initiatives, University of Victoria, 6:2.

World Commission on Dams (2000) *Dams and Development: A New Framework for Decision-Making*. www.dams.org (accessed 6 April 2001).

Wright, G. (1980) *Moralism and the Model Home*. Chicago: The University of Chicago Press.

Wright, J., et al. (1979) *After the Clean-Up: Long-Range Effects of Natural Disasters*. Beverley Hills, Calif.: Sage Publications.

Wright, R. (1992) *Stolen Continents: The "New World" Through Indian Eyes*. Toronto: Penguin.

Wudunn, S. (1997) "Airport Devours Farmers' Fields a Bite at a Time," *Globe and Mail* 10 September, A11.

Yardley, J. (2000) "U.S. Panel Urges Compensation for 1921 Race-Riot Survivors," *Globe and Mail* 7 February.

York, G. (1990) *The Dispossessed: Life and Death in Native Canada*. Boston: Little, Brown.

Young, M., et al. (1981) *Report from Hackney*. London: Policy Studies Institute.

Zich, A. (1997) "China's Three Gorges: Before the Flood," *National Geographic* September: 2–33.

Zonn, L. (1983) "Home." Paper presented at the annual meeting of the Association of American Geographers. Denver, Colo., April, Photocopy.

Index